T0290984

Space Micropropulsion
for Nanosatellites

Space Micropropulsion for Nanosatellites
Progress, Challenges and Future

Edited by

Kean How Cheah, PhD

*Assistant Professor, School of Aerospace,
Faculty of Science and Engineering,
University of Nottingham Ningbo China,
Ningbo, Zhejiang, China*

ELSEVIER

Elsevier
Radarweg 29, PO Box 211, 1000 AE Amsterdam, Netherlands
The Boulevard, Langford Lane, Kidlington, Oxford OX5 1GB, United Kingdom
50 Hampshire Street, 5th Floor, Cambridge, MA 02139, United States

Notices
Knowledge and best practice in this field are constantly changing. As new research and experience broaden our understanding, changes in research methods, professional practices, or medical treatment may become necessary.

Practitioners and researchers must always rely on their own experience and knowledge in evaluating and using any information, methods, compounds, or experiments described herein. In using such information or methods they should be mindful of their own safety and the safety of others, including parties for whom they have a professional responsibility.

To the fullest extent of the law, neither the Publisher nor the authors, contributors, or editors, assume any liability for any injury and/or damage to persons or property as a matter of products liability, negligence or otherwise, or from any use or operation of any methods, products, instructions, or ideas contained in the material herein.

Library of Congress Cataloging-in-Publication Data
A catalog record for this book is available from the Library of Congress

British Library Cataloguing-in-Publication Data
A catalogue record for this book is available from the British Library

ISBN: 978-0-12-819037-1

For information on all Elsevier publications visit our website at
https://www.elsevier.com/books-and-journals

Publisher: Matthew Deans
Acquisitions Editor: Brian Guerin
Editorial Project Manager: Emily Thomson
Production Project Manager: Sreejith Viswanathan
Cover Designer: Matthew Limbert

Typeset by TNQ Technologies

Contents

Contributors

Angelo Cervone, PhD
Assistant Professor, Aerospace Engineering Faculty, Delft University of Technology, Delft, the Netherlands

Wai Siong Chai, PhD
Post-doctoral Researcher, School of Mechanical Engineering and Automation, Harbin Institute of Technology, Shenzhen, Guangdong, China

Kean How Cheah, PhD
Assistant Professor, School of Aerospace, Faculty of Science and Engineering, University of Nottingham Ningbo China, Ningbo, Zhejiang, China

Luigi T. DeLuca, PhD
Professor, Institute of Space Propulsion, School of Chemical Engineering, Nanjing University of Science and Technology, Nanjing, Jiangsu, China; Space Propulsion Laboratory, Department of Aerospace Science and Technology, Politecnico di Milano, Milano, Italy

Chaggai Ganani, MSc
Graduate Student, Aerospace Engineering Faculty, Delft University of Technology, Delft, the Netherlands

Dadui Cordeiro Guerrieri, PhD
Aerospace Engineering Faculty, Delft University of Technology, Delft, the Netherlands

Zhang He, MS
Senior Engineer, Institute of Space Propulsion, School of Chemical Engineering, Nanjing University of Science and Technology, Nanjing, Jiangsu, China; The Key Laboratory of Nano-micro Energetic Devices of Ministry of Industry and Information Technology, Nanjing University of Science and Technology, Nanjing, Jiangsu, China

Akira Kakami, PhD
Associate Professor, Department of Aeronautics and Astronautics, Tokyo Metropolitan University, Hino, Tokyo, Japan

Toshiyuki Katsumi, PhD
Associate Professor, Department of Mechanical Engineering, Nagaoka University of Technology, Nagaoka, Niigata, Japan

Kai Seng Koh, PhD
Assistant Professor, School of Engineering and Physical Sciences, Heriot-Watt University Malaysia, Putrajaya, Malaysia

Kristina Lemmer, PhD
Associate Professor, Department of Mechanical and Aerospace Engineering, Western Michigan University, Kalamazoo, MI, United States

Fiona Leverone, PhD
Aerospace Engineering Faculty, Delft University of Technology, Delft, the Netherlands

Yimeng Li, BEng
Graduate Student, School of Aerospace, University of Nottingham Ningbo China, Ningbo, Zhejiang, China

Chengbo Ru, PhD
Lecturer, Institute of Space Propulsion, School of Chemical Engineering, Nanjing University of Science and Technology, Nanjing, Jiangsu, China; The Key Laboratory of Nano-micro Energetic Devices of Ministry of Industry and Information Technology, Nanjing University of Science and Technology, Nanjing, Jiangsu, China

Ruiqi Shen, PhD
Professor, Institute of Space Propulsion, School of Chemical Engineering, Nanjing University of Science and Technology, Nanjing, Jiangsu, China; The Key Laboratory of Nano-micro Energetic Devices of Ministry of Industry and Information Technology, Nanjing University of Science and Technology, Nanjing, Jiangsu, China

Marsil de Athayde Costa e Silva, PhD
Aerospace Engineering Faculty, Delft University of Technology, Delft, the Netherlands

Xiaoyong Wang, MS
Research Associate, Institute of Space Propulsion, School of Chemical Engineering, Nanjing University of Science and Technology, Nanjing, Jiangsu, China; The Key Laboratory of Nano-micro Energetic Devices of Ministry of Industry and Information Technology, Nanjing University of Science and Technology, Nanjing, Jiangsu, China

Gabe Xu, PhD
Associate Professor, Mechanical and Aerospace Engineering, University of Alabama in Huntsville, Huntsville, AL, United States

Yinghua Ye, PhD
Professor, Institute of Space Propulsion, School of Chemical Engineering, Nanjing University of Science and Technology, Nanjing, Jiangsu, China; The Key Laboratory of Nano-micro Energetic Devices of Ministry of Industry and Information Technology, Nanjing University of Science and Technology, Nanjing, Jiangsu, China

Introduction

Emerging of nanosatellites

Kean How Cheah, PhD

Assistant Professor, School of Aerospace, Faculty of Science and Engineering, University of Nottingham Ningbo China, Ningbo, Zhejiang, China

1.1 Philosophy of micro- and nanosatellites

By the broadest definition, a satellite is a natural or artificial (man-made) body that orbits another body in outer space. For example, the moon is the natural satellite that is orbiting the Earth while the Earth is the natural satellite of the Sun. An artificial satellite (simply known as satellite thereafter) is a man-made object launched into space to orbit a predetermined body, for exampe, Earth, Moon, Mars as well as the Sun. Satellites are commonly categorized according to their physical size as shown in Table 1.1.

The first Earth-orbiting artificial satellite is the Soviet Union built Sputnik-1. It was sent into space on October 4, 1957, which marked the start of space exploration in human history [1]. Microsatellite has shown their existence at the very early stage of spaceflight. The Sputnik-1 is 83.6 kg in mass (Fig. 1.1A), while the first US-built satellite that reached the orbit, the Explorer 1, has a mass of 14 kg only (Fig. 1.1B) [2]. It is noteworthy that it was not the lack of technical know-how to build larger satellites but the less capability of launch vehicles of the time has limited the putting of larger satellites into space. As the capability of launch vehicles increased steadily, we have witnessed a shift in paradigm to "bigger is better," as evidenced by the leap in satellite mass of Sputnik-2 to 508 kg and Sputnik-3 to a whopping mass of 1327 kg. As the space race between the United States and the Sovient Union

Table 1.1 Classification of satellites.

Class	Mass (kg)
Large satellite	>1000
Small satellite	500−1000
Minisatellite	100−500
Microsatellite	10−100
Nanosatellite	1−10
Picosatellite	0.1−1
Femtosatellite	<0.1

Space Micropropulsion for Nanosatellites. https://doi.org/10.1016/B978-0-12-819037-1.00001-3

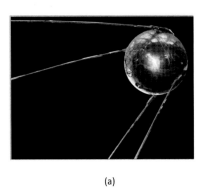

(a)   (b)

FIGURE 1.1

(A) Sputnik-1 with the radio transmitting antennas [1]; (B) Photo of William H. Pickering, James A. Van Allen, and Wernher von Braun (from left to right) carrying a full-scale model of the Explorer 1 satellite.

Reprint with permission from F. McDonald, J.E. Naugle, Discovering Earth's radiation belts: remembering explorer 1 and 3, Eos, Trans. Am. Geophys. Union 89 (39) (2008) 361–363.

intensified in the 1960s, the two space nations were competing to build larger, heavier, and more complex satellites and launch into space in the next 30 years.

Such a trend was worrying the space community. The mission cost has since ballooned as a large aerospace organization or institution with a dedicated team of advanced technical expertise is required to not only design and develop, but also the subsequent management and operation of these sophisticated large satellites. The huge investment needed in a large satellite has practically limited the launch opportunity. For a large satellite with a complex mission, the development lifecycle could easily take more than 10 years as a series of comprehensive, stringent, and thus lengthy quality assurance and testing processes is necessary to ensure the reliability and robustness of the satellite before it is ready for launch. It was becoming a norm that compromised solutions were increasingly made to integrate diverse and incompatible functions into a large satellite platform [3] as flight opportunity is getting scarce. As a result, the compromised design could be poor and inefficient. The high mission cost coupled with the limited flight opportunity has provided room for conservatism to creep in. When any failure in a system could potentially ruin the entire costly mission, innovative and risky ideas are no longer preferable and encouraged.

Small satellites—Big future, title of 2010 Appleton Lecture by Prof. Sir Martin Sweeting at The Institute of Engineering and Technology on Jan 19, 2010

Realizing these demerits, micro- and nanosatellites have regained the attention of the space community. Functionalities of a large satellite could be distributed into several smaller satellites, which are less complex and economic to develop in a

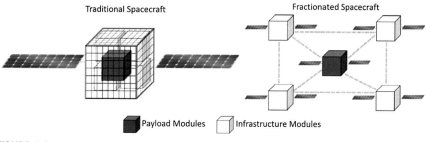

Traditional Spacecraft   Fractionated Spacecraft

■ Payload Modules   ⌂ Infrastructure Modules

FIGURE 1.2

Comparison of traditional spacecraft and fractionated spacecraft.

Reprint with permission from A. Poghosyan, A. Golkar, CubeSat evolution: analyzing CubeSat capabilities for conducting science missions, Prog. Aero. Sci. 88 (2017) 59–83.

shorter period (Fig. 1.2) [4]. A malfunction in any smaller satellite in space can be readily replaced within a reasonable timeframe. Such "smaller, faster, better, cheaper" approach has provided NASA a cost-effective and therefore sustainable way in carrying out their future near-Earth scientific missions as well as interplanetary explorations. For example, NASA has planned to send at least a spacecraft to Mars during each Earth—Mars launch window opportunity in the future.

Another key catalyst to the adoption of "smaller" satellite approach is the rapid advancement in microelectronics, in particular the microelectro-mechanical system technology. In 1965, Gordon Moore has first postulated the number of components in an integrated circuit to double every year in the coming decade (famously known as Moore's law). He then revisited the proposal in 1975 and revised the forecast to a sustained rate of increase (doubling every year) until 1980 before it is reduced to slower a rate of doubling every 2 years [5]. Even with the reduced rate of increase, the pace of technological advancement in microelectronic is still considered "fast" if not "rapid" for the space industry. A large satellite mission that takes 10 years to complete would see its technological level in electronic systems "obsolete" by the time it is ready for launch. Such development in the consumer electronics industry has very positive motivation and influence for the space industry to reform from its traditionally cautious and rather slow pace mode of operation to one that is responsive and innovative.

A design-to-cost approach in developing a satellite, which imposes strict control over cost and schedule, was introduced at the third United Nations Conference on the Exploration and Peaceful Uses of Outer Space (UNISPACE III), Vienna, Austria in 1999. The approach was later pioneered by Surrey Satellite Technology Ltd. to contain the cost overrun, a persisting issue encountered in previous large satellite missions that adopted a design-to-capability approach [6]. The new approach encourages the utilization of up-to-date microelectronic technologies and is particularly suited for micro- and nanosatellites with relatively short mission lifetime, which translates into relatively less stringent requirements on the system components. After carefully screening and evaluation toward their susceptibility to the adverse effect of space environments, especially the highly energetic galactic rays and the trapped particles in the Van Allen radiation belt, the commercial-off-the-shelf (COTS) microelectronic components could be widely used in micro- and nano-satellites. This reduced the overall mission cost significantly without compromising the performance of the satellites in orbit. To date, COTS components with reasonably low cost, consumed less electrical power and spatial volume have proven capable to perform reliably in the low earth orbits [7].

Apart from the adoption of COTS components, another key factor to the resurgence of "smaller" satellite is the changing of management approach to one that draws inspiration from the IT industry management model, which employs small but responsive team to execute the project swiftly. This is in stark contrast to the conventional large aerospace organizations with a large number of staffs, layered management structures, and procedures. Specifically, the eleven (11) characteristics of such small and agile satellite teams have been outlined. Readers who would like to know more are encouraged to read the review on how modern small satellites change the economics of space [7]. The philosophy of micro- and nanosatellite becomes attractive for established space agencies as they now have more mission opportunities to demonstrate, test, and mature the novel technological ideas in space.

For the past few decades, space exploration has been dominated by the developed or relatively rich developing nations, such as the United States, China, Russia, countries of European Union, Japan, India, and Canada, primarily because of the high demands in advanced technological and economic resources to initiate a space program. The privilege to space has granted the people in those countries to enjoy better living, economic, and social standards, brought by the satellite technologies, such as weather forecast, communication, remote sensing, navigation as well as homeland security.

For developing and emerging countries, most of the resources are utilized for near-term and urgent needs, such as food, clean water, health care, education, and infrastructure. However, it is necessary to invest in human capitals, especially through strategic and advanced technological sectors, which is critical for the long-term development and prosperity of a country. To this end, micro- and nanosatellite program is particularly beneficial for these countries, by providing a unique and economically affordable pathway to expand their intellectual capitals [8] through scientific investigation and technological capacity building in the final frontier. The reduced mission complexity has allowed these countries to initiate their own space program, train their scientists and engineers to develop own communication, earth observation or defense security satellites, and eventually launch into orbit. Performance of their micro- and nanosatellites may not be comparable to those larger satellites but they have direct control and access to their satellites without relying on the major satellite service providers.

Although there are various benefits and advantages offered by micro- and nanosatellites as discussed previously, large satellites still have a unique role in the space industry and remain the preferred satellite form factor for certain applications. For example, GEO communication applications prefer huge and powerful satellites owing to their multiple payloads carrying capability to provide adequate bandwidth for the area of coverage.

1.2 The birth of CubeSats

After years of continuous development, the capability of micro- and nanosatellites has been steadily improving in terms of platform design, payload performance, and ground control management. Around the year 2000, these "smaller" satellites have accumulated sufficient capability and evolved from the early days of demonstration purpose into useful utility missions, in particular earth observation of commercial value [7]. Specifically, the improvement in attitude control, precise pointing, onboard data storage, and data downlink combined with the advance in imaging sensor technology has increased the resolution of images captured in low earth orbits. For instance, the three-axis stabilized TopSat, a 120 kg satellite launched in 2005, was able to capture images with resolution down to 2.8 and 5.6 m of ground sample distance for panchromatic and multispectral images, respectively. The captured images could be downlinked to a ground station within a few minutes using its X-band transponder at 11 Mb/s [9].

While microsatellites (<100 kg) have reduced the mission cost significantly, they are still considered capital intensive for educational purposes. Establishing a comprehensive educational training program, which covers all stages of satellite missions, that is, from design, manufacturing, assembly, testing, launch, and finally operate the satellite, is essential to provide valuable hands-on experiences in nurturing the existing students into competent scientists and engineers to meet the growing demand in the space industry.

Recognizing the need for science and engineering students to get involved in a real satellite mission from early on, the idea of employing much smaller satellites, between nano- and picosatellite class, was considered. In 1999, Jordi Puig-Suari (from California Polytechnic State University, Fig. 1.3A) and Bob Twiggs (from Stanford University, Fig. 1.3B) have collaborated and initiated a picosatellite project, which was supposed to launch on a Russian rocket [10]. However, the launch opportunity was canceled unexpectedly. This has prompted them to start thinking to transform the pico-class satellite into a standard to extend the reach of such tiny satellite among the academic communities. With a mass of less than 1 kg, the physical size of a picosatellite is small. To harvest the maximum solar power and convert it into useful electricity, one needs to mount as many solar cells as possible on the surfaces of a picosatellite, such consideration leads to the setting of the picosatellite in a cube shape. The idea eventually evolved into a standardized pico-class satellite in a form factor of 10 cm cube (which can hold 1 kg of water), and thus known as CubeSat [11]. The inception of CubeSat has attracted tremendous interest from universities around the world, with almost all major universities have had their

(a)   (b)

FIGURE 1.3

(A) Prof. Jordi Puig-Suari, who is holding a 1U CubeSat; (B) Prof. Bob Twiggs.

Reprint with permission from R.A. Deepak, R.J. Twiggs, Thinking out of the box: space science beyond the
CubeSat, J. Small Satell. 1 (1) (2012) 3–7.

own CubeSat program. The rapid advances in the fields of microelectronics, tele-communications, materials, and instrumentations have accelerated the development of CubeSat-related technologies and reduced the cost further. Now, it is not just affordable for established and big universities, but smaller universities, even high school students could access space via a CubeSat program.

CubeSats are classified as pico- to nanoclass of satellites, which were built according to a specific set of standards such as shape, size, and weight as outlined in the Cubesat design specifications (CDS). A standard one CubeSat unit (1U) is referring to a $10 \times 10 \times 10$ cm cube with a mass of up to 2 kg (an increase from 1.33 kg, as stated in the latest CDS Rev. 14). Throughout the years of development, CubeSats have grown into larger sizes, such as 1.5, 2, 3, 6, and 12U (even size as large as 24U has been proposed) to meet the increasing demand of the mission on advanced functionality and capability of the satellites. Examples of the mechanical layout of a 1, 2, 3, and 6U CubeSat size are shown and compared in Fig. 1.4 [12].

Standardizing the specifications of the satellites offers multiple advantages. Smaller companies can develop individual CubeSat subsystems, mass-produce us-ing COTS components, and supply to different CubeSat developers, who later stack and assemble the subsystems into a highly integrated CubeSat according to the mission requirements (Fig. 1.5) [13]. Such modular approach of satellite develop-ment permits batch-production, in which the economy of scale would reduce the overall cost, as witnessed in the consumer electronic products, such as PCs, smart-phones, and other gadgets. The setting up of standards is also promoting the sharing and exchange of information among the CubeSat community as a common "lan-guage" has been widely used in communication. Since then, many CubeSat-

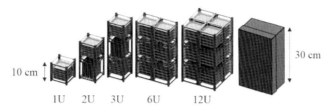

FIGURE 1.4

Volume of a 1, 2, 3, and 6U CubeSat (start from left).

Reprint with permission from P. Machuca, J.P. Sánchez, S. Greenland, Asteroid flyby opportunities using semi-autonomous CubeSats: mission design and science opportunities, Planet. Space Sci. 165 (2019) 179–193.

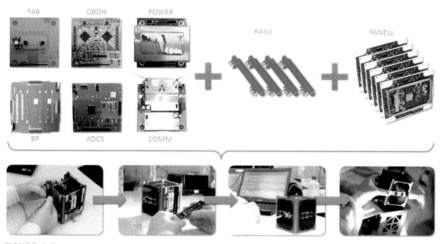

FIGURE 1.5

Modular design architecture allows integration of subsystems/components from different developers into a CubeSat.

Reprint with permission from K. Schilling, Perspectives for miniaturized, distributed, networked cooperating systems for space exploration, Robot. Autonom. Syst. 90 (2017) 118–124.

related conferences, workshops, and exhibitions have been hosted by various organizations for the stakeholders to engage and interact. Furthermore, the standardized physical dimension of CubeSats facilitates the general transportation and deployment into space as a common set of procedures and systems could be used, independent of CubeSat developers.

1.3 Launching of CubeSats

For most of the rocket launches, there is usually some excess in launch capacity after integrating the primary satellite, allowing the possible inclusion of small secondary

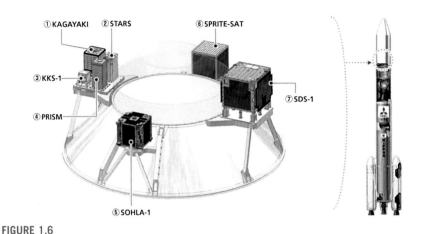

FIGURE 1.6

Piggyback satellites on H-IIA launch vehicle.

Reprint with permission from JAXA; Rocket image: Courtesy of Mitsubishi Heavy Industries Ltd.

payloads. In the early days, most of the CubeSats were launched into orbit via such ridesharing launches as secondary payload (or called piggyback satellite), sharing the ride with a primary payload (Fig. 1.6). While the associated launching cost is low, it is subjected to requirements and restrictions imposed by the launch service provider and the primary satellite owner. For example, the piggyback satellite owner has no control over the launch date, and on some occasions, needs to deliver the satellite to the launch site for integration with the launch vehicle weeks before the actual launch date to accommodate the schedule of the primary satellite. In addition, the primary mission always determines the final altitude for orbit insertion. In most cases, the deployment of CubeSats is only carried out once the primary satellite has been separated from the launch vehicle, clearing the potential risk of in-orbit collision.

With the increasing demand in CubeSat launching, there are dedicated rideshare launches, where a combination of multiple small- and/or CubeSats are launched together. As of January 2021, SpaceX holds the record with 143 satellites of different sizes launched on a single rocket beating the previous record by Indian's PSLV C-37 launch of 103 satellites. Such mode of ridesharing offers the riding satellites more equal partnership, and a therefore fairer share of control over the final altitude (as opposed to the domination by the primary satellite), where deployments of satellites at multiple altitudes are possible, as long as it is within the capability of the launch vehicle. Although dedicated rideshare launches increase the launch opportunities for CubeSats, the managing of diverse requirements from different stakeholders, that is, launch service providers and satellite developers of each ridesharing satellite, as well as the logistic arrangement, is a challenging task [7]. Currently, NASA and ESA are coordinating and supporting the launch of CubeSats developed by universities, research institutions, and schools through the CubeSat Launch Initiative and Fly Your Satellite Program, respectively.

Since 2014, there is a new option to launch the CubeSats into orbit. The CubeSats could be sent as part of the cargo on a resupply mission to the International Space Station (ISS), where the CubeSats are launched into space using a dedicated deployer. JAXA has developed the first of this kind deployer, called the Japanese Experimental Module (JEM) Small Satellite Orbital Deployer (J-SSOD) to deploy the CubeSats, up to 6U in form factor, from the ISS [14].

In a typical deployment procedure, the CubeSats are preloaded into a dispenser, known as Satellite Install Case (each can hold up to 3 units of CubeSat), and launched as cargo to ISS. Once the cargo reaches the ISS, Satellite Install Cases are transferred into the Japanese KIBO module, where a crewmember (an astronaut) will load them onto J-SSOD, which is equipped with an electrical box and separation mechanism for deployment. The J-SSOD assembly (with a maximum of two Satellite Install Cases) is then attached to the multipurpose experiment platform (MPEP) and is slid out from the KIBO's airlock to the space environment. Subsequently, the KIBO's robotic arm Japanese Experiment Module Remote Manipulator System (JEMRMS) grasps the MPEP assembly and transfers it to the deployment point (Fig. 1.7A), which is usually in the direction opposing the travel direction of ISS and pointed to nadir-aft 45 degrees. In the final step, the spring on Satellite Install Case jettisons the CubeSats into space sequentially (Fig. 1.7B) [15]. The first J-SSOD deployment was successfully carried out on October 4, 2012, in a two-time deployment. In the first deployment, it was operated by JAXA astronaut Akihiko Hoshide on orbit to release the first two CubeSats (WE WISH and RAIKO). This was followed by the second deployment operated by the mission control on the ground to release the remaining three CubeSats (FITSAT-1, F-1, and TechEd-Sat). Soon after the successful deployments of CubeSats from ISS, NanoRacks LLC (based in Houston, US) has developed the first commercial deployer, sharing the same facilities and deployment procedures with J-SSOD. The Nanoracks CubeSat Deployer (NRCSD) can hold up to 6 units of CubeSats while the upgraded NRCSD DoubleWide has doubled the capacity to 12 units of CubeSats.

On a very rare occasion, the Peruvian Chasqui 1 cubesat was "launched" by a Russian cosmonaut by tossing the tiny satellite from ISS by hand during a spacewalk in 2014.

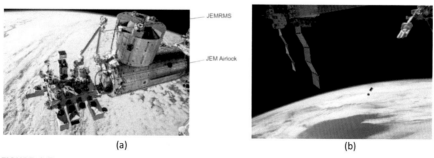

(a)   (b)

FIGURE 1.7

(A) Location of JEMRMS and JEM Airlock in KIBO lab; (B) Deployment of CubeSats.

Reprint with permission from Y. Nogawa, S. Imai, 24 - Launch from the ISS, in: C. Cappelletti, S. Battistini, B.K. Malphrus (Eds.), Cubesat Handbook, 2021, Academic Press, pp. 445–454.

1.4 First CubeSats

The first group of CubeSats was launched on June 30, 2003, with a Rokot rocket from Plesetsk, Russia. The mission has put six CubeSats into a sun-synchronous orbit: the Danish AAU CubeSat and DTUSat, the Canadian CanX-1, the US Quake-Sat, and the Japanese XI-IV and CUTE-1.

AAU CubeSat was built by students from Aalborg University, Denmark. The students had initially intended to conduct an earth observation mission but later found challenging to execute in a 1U CubeSat [16]. It was then decided that the CubeSat would carry a CMOS camera of 1.3 megapixels in 24-bit colors with a resolution of 150×120 m and used this opportunity as a technology evaluation preparing for future missions in CubeSat platform. Unfortunately, the mission was short-lived, and it lasted for about two and a half months before the battery lost all its capacity with only basic beacon data signal received on the ground. Postanalysis has suggested that the transmitted signal was too weak to establish an effective data link for communication and download of housekeeping data.

Another Danish CubeSat, the DTUSat was designed and built by students from the Technical University of Denmark. The primary payload of the CubeSat is an electrodynamic tether, which was designed to employ a 450-m bare copper wire to produce force upon interacting with the Earth's magnetic field, with the potential application in object removal from space. It also carried a camera and a radio test broadcaster as secondary payloads. Nevertheless, the communication link was not able to establish with the CubeSat after launch.

Canadian Advanced Nanosatellite eXperiment (CanX) is a Canadian picosatellite program by Space Flight Laboratory at the University of Toronto Institute for Aerospace Studies. CanX-1 is the first picosatellite built by the students to demonstrate and verify the functionalities of a small but highly capable satellite, and thus it carried a number of payloads and experimental subsystems, which included CMOS imagers, high performance ARM7 onboard computer, GPS receiver, and active magnetic attitude control system [17]. Unfortunately, the CanX-1 suffered a similar fate as DTUSat and was not able to establish communication with the ground station after launch.

QuakeSat is a 3U CubeSat built by the Space Systems Development Laboratory at Stanford University and a technology company named Quake Finder in Palo Alto to detect the early signal of an earthquake. The payload is an extremely low frequency (ELF) magnetometer, which detects and collects the ELF magnetic signal data from space for the prediction of potential earthquake activity. The sensitive ELF magnetometer was placed on a boom, which was deployed for 0.701 m from the main CubeSat bus to shield the magnetometer from magnetic interference generated by the onboard electronics. As the first 3U CubeSat in history, the QuakeSat mission was highly successful and had operated for 1.5 years, though the design life is only 1 year. It has collected unique signals before and after a few large earthquakes, such as San Simeon CA, South Island NZ, and Kazakhstan-Xinjiang Border Region earthquakes [18].

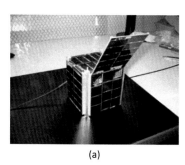
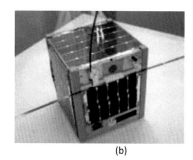

(a)   (b)

FIGURE 1.8

The first 3U CubeSat—(A) CUTE-1 flight model. (B) XI-IV CubeSat .

(A) Reprint with permission from H. Ashida, et al., Design of Tokyo Tech nano-satellite cute-1.7+APD II and its operation, Acta Astronaut. 66 (9) (2010) 1412–1424; Reprint with permission from S. Nakasuka, et al., Evolution from education to practical use in University of Tokyo's nano-satellite activities, Acta Astronaut. 66 (7) (2010) 1099–1105.

Cubical Tokyo Tech Engineering Satellite-I (CUTE-1) is a student-led 1U Cube-Sat program developed in the Laboratory for Space Systems of the Tokyo Institute of Technology [19]. The CubeSat has three mission objectives: (1) to demonstrate communication using the CW- and FM-telemetry interchangeably according to the control command from the ground station, (2) to collect the housekeeping data from different onboard sensors, such as gyroscope, accelerometers, thermometers, and sun sensor, and (3) to demonstrate the deployment mechanism for solar panel (Fig. 1.8A).

X-factor Investigator IV (XI-VI) is a 1U CubeSat developed by the Intelligent Space Systems Laboratory at the University of Tokyo [20]. There are two main mission objectives: (1) to support space engineering education and (2) to demonstrate and test a picosatellite bus system, which used COTS components extensively (except the solar cells) in orbit. In addition, the CubeSat has an advanced level mission, which employed a CMOS camera to capture the images of Earth and downlink them to the ground station. It is noteworthy that both Japanese CubeSats have survived the space environment and continuously operating into 2021, an astonishing 18 years in orbit. The world's longest operation of CubeSat to date.

1.5 CubeSats for scientific missions and commercialization

For the next 10 years after the launch of the first CubeSats, the CubeSats were still primarily used as an educational tool to train university students or engineers as well as providing a low-cost platform for new technology demonstration in space, which can be typically executed within 1−2 years. Such trend has changed since 2014 with the majority of the new launches were for scientific or commercial missions [7], developed by not only small and startup companies, but also large space agencies such as NASA and ESA, who have started their own CubeSat programs for advanced scientific missions of the potentially high value of the return.

The evolution of CubeSats has been recently analyzed to evaluate their capabilities in conducting high-quality scientific missions. Over 130 CubeSat missions were selected based on the primary mission objectives and grouped into six categories: (1) Earth science and spaceborne applications, (2) Deep space exploration, (3) Heliophysics: Space weather, (4) Astrophysics, (5) Spaceborne in situ Laboratory, and (6) Technology demonstration. Readers who are interested to know more about the development in this area are encouraged to read the review [4].

In 2017, the number of CubeSats deployed in a dedicated ridesharing launch has broken the 100 mark for the first time. 101 CubeSats were launched on the Indian PSLV rocket, of which 88 are earth observation (EO) CubeSats developed by a commercial earth imaging company named Planet, US. The possibility to launch a large fleet of CubeSats into space at once provides the key enabling support for a radically new mission architecture, that is, constellation (or cluster) of CubeSats, which has massive commercial potential on earth observation. Planet is building the largest ever constellation of EO CubeSat with a flock of 3U CubeSats (of approximately 5 kg each, called "Doves") [21] to capture a stream of multispectral imagery on the entire Earth every day at a resolution of 3−5 m in RGB near-infrared wavelengths for environmental, humanitarian, and commercial applications. Since then, the company has continuously improved its EO technology for capturing better quality and sharp images through the new generation of "SuperDoves" CubeSats and downlinks the images to a network of ground stations. As of January 2021, the company is operating a global EO constellation of over 200 active CubeSats.

1.6 CubeSats beyond the Earth

Since the born of CubeSat in 1999, 2 decades have passed. Cubesats have evolved from a "toy," as regarded by many pessimists in the early days, into formidable and capable satellites to conduct a variety of scientific and commercial missions. The space community has started to explore if the CubeSats are capable to go beyond the Earth. There are several risks and challenges that need to be addressed before the CubeSats could be considered for deep space missions. Among the firsts is long-distance communication, which requires the development of a miniaturized deep space communication system that can fit into a CubeSat form factor. Without the protection of Earth's magnetic field, the effect of the deep space environment could be very harsh to the CubeSats as the radiation level is expected to grow much stronger and the design of the thermal control system becomes ever more challenging.

Nevertheless, risks and challenges do not prevent the CubeSat from going beyond the Earth's orbit. In contrast, the risk-taking characteristics of CubeSat have propelled the first interplanetary CubeSat, Mars Cube One (MarCO) to as far as Mars in a flyby mission in 2018 (Fig. 1.9) [22]. The mission consists of

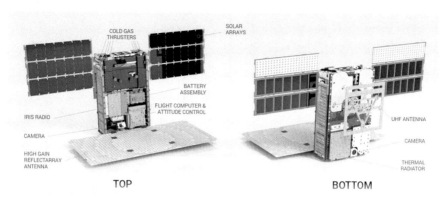

TOP   BOTTOM

FIGURE 1.9

The twin Mars Cube One (MarCO) CubeSats.

Reprint with permission from F.C. Krause, et al., Implementation of commercial Li-ion cells on the MarCO deep space CubeSats, J. Power Sources 449 (2020) 227544.

twin 6U CubeSats, MarCO-A and MarCO-B, launched together with NASA's InSight lander, with the primary objective to test the miniaturized communication and navigation technologies in the CubeSat platform. During the entry, descent, and landing of the Insight lander on November 26, 2018, the newly developed deep space communication system in MarCO had provided a real-time communication link between the lander and Deep Space Network on earth. The touchdown of InSight on Mars was confirmed by the telemetry data and the first picture was taken by the lander, which was relayed back to earth via MarCO CubeSats.

In the latest quest beyond the Earth, the next destination for CubeSat is the Moon. Artemis is a US-led international program, launched in 2017 to return humans to the Moon. The first mission, Artemis I, will be an uncrewed test flight to send the new Orion multipurpose crew vehicle, riding on the Space Launch System rocket, to a retrograde lunar orbit for 6 days. The launch is scheduled for November 2021, and NASA has exploited this unique opportunity to send thirteen (13) 6U CubeSats, as a secondary payload, to test and demonstrate a wide range of lunar exploration as well as deep space-related technologies [23]. As summarized in Table 1.2, seven (7) of the CubeSats are developed by various US government research centers, institutes, and universities. Three (3) CubeSats are from NASA's international partners and the remaining three (3) CubeSats are selected through NASA's Cube Quest Challenge competition.

1.7 The need of micropropulsion system

The small spatial volume and limited electrical power budget in CubeSats do not favor the inclusion of a propulsion system. The first CubeSat had no propulsion

Table 1.2 Micropropulsion system used in CubeSats in Artemis I mission.

Mission	Organization	Propulsion system	Model	Thrust
US Government Research Centers, institutes, and universities				
BioSentinel	NASA Ames Research Center	Cold gas	Developed by Georgia Institute of Technology	50 mN
CuSP	Southwest Research Institute	Cold gas	MiPS by VACCO	25 mN
LunaH-Map	Arizona State University	Electrostatic	BIT-3 ion thruster by Busek	1.24 mN
LunIR	Lockheed Martin Space	Electrospray	NA	NA
Lunar Flashlight	Jet Propulsion Laboratory	Monopropellant	Developed by Georgia Tech Space System Design Laboratory	100 mN
Lunar IceCube	Morehead State University	Electrostatic	BIT-3 ion thruster by Busek	1.24 mN
Near-Earth Asteroid Scout	Marshall Space Flight Center Jet Propulsion Laboratory		Use solar sail	
NASA's international partners				
ArgoMoon	Argotec Italian Space Agency	Monopropellant Cold gas	Hybrid MiPS by VACCO	100 mN 25 mN
EQUULEUS	JAXA University of Tokyo	Electrothermal	AQUARIUS by University of Tokyo	4 mN
OMOTENASHI	JAXA	Cold gas Solid propellant	MiPS by VACCO Developed JAXA	25 mN 500 N
NASA's cube quest challenge competition				
Cislunar Explorers	Cornell University	Bipropellant Cold gas	H_2/O_2 from electrolysis	1 N
Team Miles	Fluid and Reason, LLC.	Hybrid plasma and laser	ConstantQ Model H by Fluid and Reason, LLC.	5 mN
Earth Escape Explorer	University of Colorado Boulder	Does not feature a propulsion system. Use solar radiation pressure for attitude control		

system onboard. The AAU, DTU, and CanX-1 used magnetorquers for attitude control and stabilization, the QuakeSat used a permanent magnet to keep the magnetometer aligned to Earth's magnetic field, while the XV-I used a permanent magnet and a hysteresis damper to point the satellite in a geomagnetic direction.

To extend the CubeSats mission capability, a CubeSat-compatible propulsion system is crucial in providing the necessary propulsive force or impulse for both low ΔV operations, such as attitude control, orbit maintenance and correction, momentum wheel unloading, and drag compensation, and high ΔV maneuvers, such as a change in orbit altitude, constellation deployment, and formation flying, as well as deorbiting at the end of mission life to prevent the CubeSat from becoming space debris.

The need for a propulsion system (known as micropropulsion sytem thereafter) capable of producing thrust in the order of μN to mN level arises. The adoption of micropropulsion system can be traced back as early as 2000 in the SNAP-1 mission (6.4 kg) and MEPSI-1 mission (1.4 kg) in 2006. It is noteworthy that both satellites belong to the nanosatellite category, but were not built according to the CubeSat standard. The first CubeSat with a micropropulsion system is the 3U CanX-2, which was launched in 2008 and carried a cold gas microthruster.

The micropropulsion system is particularly vital to provide CubeSats with the required maneuverability and precise attitude control in sophisticated and cutting-edge missions. Of all the CubeSats in Artemis I mission, there is only one CubeSat, the Earth Escape Explorer, which need not carry a micropropulsion system but uses

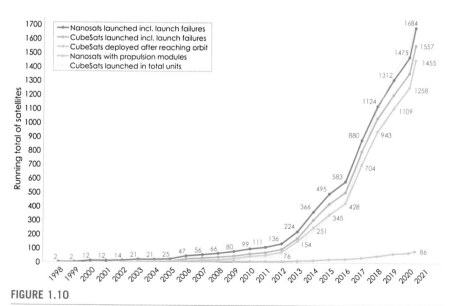

FIGURE 1.10

Total nanosatellites and CubeSats launched as of April 4, 2021.

Reprint with permission from Erik Kulu, Nanosats Database.

solar radiation pressure for attitude control. Three of the CubeSats, that is, OMOTE-NASHI, ArgoMoon, and Cislunar Explorers, even carry two different (or hybrid) micropropulsion systems to meet the highly demanding mission requirements on propulsive capability (Table 1.2).

In summary, the research and development activities in the area of micropropulsion system have intensified over the years, synchronizing with the rapid growth in the development of CubeSats. As of April 4, 2021, 86 out of 1684 nanosatellites launched (a mere 5.1%) have propulsion modules (Fig. 1.10). Nevertheless, it is expected more micropropulsion systems to achieve a higher technological readiness level and be ready for launch in the coming years. The subsequent chapters in this book will provide readers with the background and development of different types of micropropulsion systems for nanosatellites, in particular for CubeSat applications.

References

[1] Kuznetsov, V.D., Sinelnikov, V.M., Alpert, S.N., 2015. Yakov Alpert: Sputnik-1 and the first satellite ionospheric experiment. Adv. Space Res. 55 (12), 2833–2839.
[2] McDonald, F., Naugle, J.E., 2008. Discovering Earth's radiation belts: remembering explorer 1 and 3. Eos, Trans. Am. Geophys. Union 89 (39), 361–363.
[3] Myers, R., Oleson, S., Curran, F., 1994. Small satellite electric propulsion options. In: Intersociety Energy Conversion Engineering Conference. American Institute of Aeronautics and Astronautics.
[4] Poghosyan, A., Golkar, A., 2017. CubeSat evolution: analyzing CubeSat capabilities for conducting science missions. Prog. Aero. Sci. 88, 59–83.
[5] Moore, G.E., 2006. Progress in digital integrated electronics. IEEE Solid-State Circuits Soc. Newslett. 11 (3), 36–37.
[6] Kramer, H.J., Cracknell, A.P., 2008. An overview of small satellites in remote sensing. Int. J. Rem. Sens. 29 (15), 4285–4337.
[7] Sweeting, M.N., 2018. Modern small satellites-changing the economics of space. Proc. IEEE 106 (3), 343–361.
[8] Woellert, K., et al., 2011. Cubesats: cost-effective science and technology platforms for emerging and developing nations. Adv. Space Res. 47 (4), 663–684.
[9] Baxter, E., Levett, B., 2008. TopSat: lessons learned from a small satellite mission. In: Sandau, R., Röser, H.-P., Valenzuela, A. (Eds.), Small Satellites for Earth Observation. Springer Netherlands, Dordrecht, pp. 377–384.
[10] Deepak, R.A., Twiggs, R.J., 2012. Thinking out of the box: space science beyond the CubeSat. J. Small Satell. 1 (1), 3–7.
[11] Introduction: the history of the CubeSat by Bob Twiggs and Jordi Puig-Suari. In: Cappelletti, C., Battistini, S., Malphrus, B.K. (Eds.), 2021. Cubesat Handbook. Academic Press, pp. xxi–xxviii.
[12] Machuca, P., Sánchez, J.P., Greenland, S., 2019. Asteroid flyby opportunities using semi-autonomous CubeSats: mission design and science opportunities. Planet. Space Sci. 165, 179–193.
[13] Schilling, K., 2017. Perspectives for miniaturized, distributed, networked cooperating systems for space exploration. Robot. Autonom. Syst. 90, 118–124.

[14] Fite, N.D., 2021. 22 - Deployers. In: Cappelletti, C., Battistini, S., Malphrus, B.K. (Eds.), Cubesat Handbook. Academic Press, pp. 415−429.

[15] Nogawa, Y., Imai, S., 2021. 24 - Launch from the ISS. In: Cappelletti, C., Battistini, S., Malphrus, B.K. (Eds.), Cubesat Handbook. Academic Press, pp. 445−454.

[16] Alminde, L., et al., 2003. Educational value and lessons learned from the AAU-CubeSat project. In: International Conference on Recent Advances in Space Technologies, vol. 2003.

[17] Wells, G.J., Stras, L., Jeans, T., 2002. Canada's smallest satellite: the Canadian advanced nanospace eXperiment (CanX-1). In: 1st AMSAT Space Symposium and Annual Meeting (Toronto, Canada).

[18] Flagg, S., et al., 2004. Using Nanosats as a proof of concept for space science missions: QuakeSat as an operational example. In: 18th AIAA/USU Conference on Small Satellites (Logan, Utah, US).

[19] Nakaya, K., et al., 2003. Tokyo Tech CubeSat: CUTE-I - design & development of flight model and future plan. In: 21st International Communications Satellite Systems Conference and Exhibit. American Institute of Aeronautics and Astronautics.

[20] Funase, R., et al., 2019. On-orbit operation results of the world's first CubeSat XI-IV - lessons learned from its successful 15-years space flight. In: 33rd Annual AIAA/USU Conference on Small Satellites (Logan, Utah, US).

[21] Zimmerman, R., et al., 2017. Commissioning the world's largest satellite constellation. In: 31st Annual AIAA/USU Conference on Small Satellites (Logan, Utah, US).

[22] Krause, F.C., et al., 2020. Implementation of commercial Li-ion cells on the MarCO deep space CubeSats. J. Power Sources 449, 227544.

[23] Robinson, K.F., et al., 2020. Space launch system Artemis I CubeSats: SmallSat vanguards of exploration, science and technology. In: 32nd Annual AIAA/USU Conference on Small Satellites (Logan, Utah, US).

Chemical micropropulsions

Cold gas microthruster

2

Kean How Cheah, PhD [1], Chaggai Ganani, MSc [2], Kristina Lemmer, PhD [3], Angelo Cervone, PhD [4]

[1]*Assistant Professor, School of Aerospace, Faculty of Science and Engineering, University of Nottingham Ningbo China, Ningbo, Zhejiang, China;* [2]*Graduate Student, Aerospace Engineering Faculty, Delft University of Technology, Delft, the Netherlands;* [3]*Associate Professor, Department of Mechanical and Aerospace Engineering, Western Michigan University, Kalamazoo, MI, United States;* [4]*Assistant Professor, Aerospace Engineering Faculty, Delft University of Technology, Delft, the Netherlands*

2.1 Background and principles of operation

The cold gas propulsion system is the simplest form of all space propulsions. Historically, it has been used as a reaction control system for more than 50 years. They are not only used as an in-space propulsion system for spacecraft, satellites, and interplanetary landers but also find applications in the launch vehicle. For instance, the Pegasus launch vehicle uses cold gas thrusters in controlling roll axis motion [1]. The reusable first stage rocket of Falcon 9 launch vehicle from SpaceX uses the cold gas thrusters for stabilization before falling back to Earth.

A typical cold gas propulsion system consists of a high-pressure gas tank, a pressure regulator, one or more nozzles, an electrical valve for each nozzle, and slots for filling and venting of the gas, as schematically illustrated in (Fig. 2.1).

The operational principle of the cold gas propulsion system is rather simple. The gas propellant stored at high pressure is vaporized or sublimated via a phase change process into gas, which is vented through a valve into a converging-diverging (CD) nozzle, accelerating the gas flow into high velocity to produce thrust.

The gas flow through the nozzle is primarily driven by the pressure difference between the settling chamber, P_0, and the ambient atmosphere, P_b, where the gas is discharged. The gas tank pressure is high after the loading of gaseous propellant

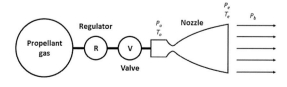

FIGURE 2.1

Schematic of a typical cold gas propulsion system.

Space Micropropulsion for Nanosatellites. https://doi.org/10.1016/B978-0-12-819037-1.00006-2

23

and reduces over time as it is consumed. If a stable and constant thrust is required, a pressure regulator is installed to stabilize the pressure of gas flowing out from the propellant tank to yield a consistent pressure difference.

Some possible gaseous propellants considered for early cold gas propulsion systems and their properties are listed in Table 2.1. While the use of hydrogen and helium gas as propellants provides higher specific impulses, the low density of these gases requires a large and heavy storage tank, which neutralizes the advantage in performance. On the other end of the spectrum, the heavier gas, such as Argon and Krypton, has low specific impulse. In some cases, these gaseous propellants are heated to improve the specific impulse or reduce the tank size, at the expense of a more complex design.

Cold gas propulsion with low system complexity is ideal for missions with relatively simple propulsive requirements. The simplicity of the system reduces the failure risk significantly as well, making it suitable for missions that require a robust and reliable propulsive system. Another advantage of cold gas propulsion is the free from contamination issue as a result of using inert and benign gaseous propellant. This is particularly important for missions that carry instruments, such as sensitive optical payload and mirrors, with stringent requirements on particle deposition and contamination.

Valve leakage remains the major limitation and concern for a cold gas propulsion system. Before the gaseous propellants flow into the valve, they must be adequately filtered to remove any microscopic contaminants, which could deposit on the valve seat, creating small voids where the leakage occurs. While the voids could be extremely small and not significant for liquid propellant-based systems, it is catastrophic for cold gas propulsion systems where the propellant viscosity is low and the pressure is high. In practice, an additional propellant of approximately 20% is normally filled into the tank as a design margin to compensate for potential leakage.

While cold gas propulsion has a rich heritage, most of the systems were not initially designed for nanosatellite applications. Although propellant tank and nozzle could be conveniently miniaturized according to the available volume and mass of a

Table 2.1 Properties of selected propellants for cold gas propulsion system [1].

Propellant	Molecular mass	Density[a] (lb/ft³)	k	Theoretical specific impulse[b] (s)
Hydrogen	2.0	1.77	1.40	284
Helium	4.0	3.54	1.67	179
Nitrogen	28.0	24.7	1.40	76
Air	28.9	25.5	1.40	74
Argon	39.9	35.3	1.67	57
Krypton	83.8	74.1	1.63	50

[a] At 5000 psia and 20°C.
[b] In vacuum with a nozzle area ratio of 50:1 and temperature of 20°C.

nanosatellite, the valve requires further research and development. Conventional valve actuation consumes notably high electrical power, in the order of 1 W for holding and as high as 10 W for opening [2], which exceeds the power rating of most nanosatellites.

Typically, cold gas propulsion system produces considerably higher thrust, which is desirable for rapid orbital maneuvers, than electric propulsion systems. However, the specific impulse is rather low, less than 100 s. Considering the limited volume in nanosatellites to integrate large storage tanks, the total impulse is insufficient for high ΔV maneuvers. This limits their applications in nanosatellites primarily as an attitude control system. In addition, the leakage issue also hinders the cold gas propulsion system from implementing in long-duration missions.

2.2 Nozzle theory

Other than the hardware, that is, propellant tank, pressure regulator, and valve, the performance of a cold gas propulsion system is closely related to the design of the nozzle. Thus, the knowledge in ideal nozzle theory, which simplifies the real nozzle flow into a quasi-one-dimensional (1D) nozzle flow, is crucial for a preliminary estimation of the thruster performance. Under the ideal 1D nozzle flow condition, the following assumptions are made:

1. The working fluid is homogeneous.
2. The working fluid is gaseous. Condensed phases might exist but they are minimal and thus negligible.
3. Perfect gas law is valid.
4. There is no heat transfer across the solid boundary, that is, the flow is adiabatic.
5. The solid boundary is smooth and therefore frictional and boundary layer effects are negligible.
6. Shock wave or flow discontinuities do not occur in the nozzle flow.
7. The flow is steady, and the transient effect is negligible.
8. Exhaust gases leave the nozzle in the axial direction only.
9. Gas properties, for example, velocity, pressure, temperature, and density, are uniform in a normal cross-section of the nozzle flow.
10. Chemical equilibrium is established, that is, frozen flow.

Ideal nozzle theory applies to all chemical propulsion systems, which use a CD nozzle to accelerate the working fluid. Although cold gas propulsion is often categorized as chemical propulsion, no chemical reaction occurs. Thus, the gaseous propellant, after flowing through the opening valve, enters the nozzle as a gas phase only. Therefore, assumptions 1, 2, and 3 in the ideal nozzle theory are all valid.

Collectively, assumptions 4, 5, and 6 allow the use of isentropic relations on the nozzle flow, which is idealized as adiabatic and thermodynamically reversible. Practically, the cold gas microthruster employs a much smaller nozzle. Thus, the assumptions shall be revised and corrected to reflect the effects of miniaturization, that is, enhanced heat loss and frictional loss at the microscale, on thruster performance.

Essentially, a CD nozzle in cold gas propulsion converts internal energy of the gas stored at high pressure of P_0 and temperature of T_0, through an isentropic gas expansion process, into kinetic energy. The gas pressure and temperature drop while the gas velocity increases to Mach number, $M = 1$ at nozzle throat and expands further to supersonic flow in the diverging section.

The total thrust produced by the nozzle flow is the sum of momentum thrust and pressure thrust that originates from the difference between the pressure at nozzle exit and ambient (back) pressure. The thrust produced is thus given by

$$F = \dot{m}V_e + (P_e - P_b)A_e \tag{2.1}$$

where A_e is the area at the nozzle exit. If the backpressure, P_b, is matching the exit pressure, P_e, optimum design condition is achieved. The thrust produced is at maximum and is a function of gas mass flow rate, $\dot{m}$, and exit velocity, V_e, only, which can be determined using

$$V_e = \sqrt{\frac{2k}{k-1}RT_0\left[1 - \left(\frac{P_e}{P_0}\right)^{(k-1)/k}\right]} \tag{2.2}$$

where k is the specific heats ratio and R is the gas constant.

The mass flow through the nozzle is a function of the throat area, A_t, and chamber pressure, P_0, and it is given by

$$\dot{m} = A_t P_0 k \frac{\sqrt{\left(\frac{2}{k+1}\right)^{(k+1)/(k-1)}}}{\sqrt{kRT_0}} \tag{2.3}$$

Specific impulse is an important performance indicator used to measure the effectiveness of a propulsion system in consuming the propellant to produce thrust. It is defined as follows:

$$I_{sp} = \frac{F}{\dot{m}g} \tag{2.4}$$

A high specific impulse implies that the propulsion system consumes less amount to propellant to produce a specific amount of thrust. This is the reason for adding thermal energy into the gas propellant via heating to reduce the mass flow rate, according to Eq. (3.3), thus increasing the specific impulse.

2.3 Selection of propellant

The choice of a suitable propellant is crucial in the development of a cold gas micro-thruster. The design considerations are highly dependent on the selected propellant.

The associated design solutions, such as sizing of the propellant storage tank, total impulse, thrust level, duty cycle, and maneuver duration, for a cold gas microthruster using one specific propellant, could be vastly different from the others.

The physical and chemical properties of a few propellants selected for cold gas microthruster are summarized and given in Table 2.2.

Inert gases, which do not undergo undesired chemical reactions, are among the first propellants selected to facilitate the safety issue. Nitrogen (N_2) and argon (Ar) gas are suitable candidates due to their inert nature as well as the well-established handling protocol and process inherited from various industries. Although butane is an organic compound and classified as a hazard, it has been selected as a propellant in previous missions due to its rich heritage in the past decades [3].

Depending on the mission requirements, there are different criteria for the selection of propellant. From Newton's laws of motion, the thrust produced by a cold gas microthruster is proportional to the mass flow rate of the gas expelled from the micronozzle. Thus, propellant with high molecular weight, such as xenon (Xe), is desirable for a mission that requires a high thrust level.

Ideally, the selected propellant is compressible into a liquid to save the inherently limited spatial volume in a nanosatellite. To this end, the propellant with high liquid density is favored as it maximizes the amount of propellant stored in the storage tank. The increased amount of propellant that a nanosatellite can carry into space extends the mission lifetime as the microthruster can operate for a longer duration.

It is beneficial for a propellant to possess the self-pressurization capability, which simplifies the design of subsystems. A dedicated pressurization system could be omitted to save not only the system mass and volume but also eliminate the proneness to pump failure or leakage in the connecting pipeline. For this criteria, propellant with low critical temperature is preferred so that the vapor pressure remains sufficient for effective self-pressurization in the operational temperature range of

Table 2.2 Physical and chemical properties of commonly selected propellant for cold gas microthruster.

	Molecular weight (g/mol)	Liquid density (g/cm³)	Heat of vaporization (kJ/mol)	Critical temperature (°C)	Freezing point (°C)
Xe	131.3	3.100	12.64	16.6	−118.8
N_2	28	0.807	2.79	−146.9	−210
Ar	39.9	1.394	6.5	−122.4	−189.4
Butane	58.1	0.599	22.44	152.2	−138
SF_6	146.1	1.329	9.64	45.6	−64
R134	102	1.225	22.15	101.1	−96.7
R236fa	152	1.373	24.38	124.9	−103

the microthruster. Nevertheless, it is noteworthy that propellant with excessively low critical temperature yields high vapor pressure when the propellant is exposed to high temperature. Under this circumstance, a propellant storage tank with a thicker tank wall, thus heavier, is necessary to avoid the over-pressurization, which could be detrimental in the event of a rupture.

Owing to the limited volume available in nanosatellites, any strategy to optimize the usage of volume is highly encouraged. Compatibility of the propellant with control and power electronics offers a possibility to pack those electronic components inside the tank, which frees up more volume in exchange for additional propellant mass. Moreover, the waste heat from the electronics could be harvested to facilitate the generation of additional vapor pressure. Therefore, the propellants from the inert and noncombustible gas category are favorable.

The temperature surrounding the nanosatellite could drop rapidly beyond 0°C when the nanosatellite is at the eclipse region. Hence, the selected propellant must have a low freezing point, which could eliminate or reduce the consumption of electrical power for active thermal control.

Sulfur hexafluoride (SF_6), a nonflammable, odorless, and nontoxic gas, is commonly used as dielectric gas in the electrical industry. It is considered as a competent propellant candidate for cold gas microthruster owing to its high molecular weight and relatively high liquid density as well as the low critical temperature at 45.6°C with good self-pressurization capacity. However, it is evaluated as one of the most potent greenhouse gases, making it a less "green" candidate.

In searching for the next generation propellant, commercial refrigerants, such as R132a and R236fa, have attracted much attention recently [4,5]. Advantages of these refrigerants, such as favorable thermodynamic properties, nontoxic, compatible with most electrical and mechanical components, and high liquid density, are appealing for application in cold gas microthruster.

In certain circumstances, there are some special requirements to meet when selecting the propellant. For instance, the deployment of CubeSats from the International Space Station (ISS) has become more common. To comply with the safety regulation in ISS, the propellant used shall be nontoxic and nonhazardous. For this purpose, R236fa is the ideal candidate as it is benign and, more importantly, it can be removed from the air by the filtration system of ISS [6].

2.4 State of the art—system with flight heritage

Cold gas systems are the most mature and established propulsion technology due to their low complexity, inexpensiveness, and robustness. Thus, it is not unexpected that they are among the first propulsion technologies successfully miniaturized for nanosatellite applications. A summary of cold gas microthruster system with flight heritage is compiled and presented in Table 2.3.

Table 2.3 Summary of cold gas microthruster system with flight heritage.

Gas	Manufacturer	Thrust (mN)	Mission	Satellite mass (kg)	Year	Reference
Butane	SSTL	65	SNAP-1	6.5	2000	[7,8]
Xenon	Aerospace corporation	20	MEPSI-1	1.4	2006	[9]
SF_6	SFL	50	CanX-2	3.5	2008	[10,11]
N_2	NanoSpace	0.1–1	PRISMA	145	2010	[12,13]
N_2	TNO, U. Twente, and TU Delft	6	Delfi-n3xt	3.5	2013	[14]
SF_6	SFL	12.5–50	CanX-4/5	1.5	2014	[15]
Ar	Microspace Rapid Pte Ltd	1	POPSAT-HIP1	3.3	2014	[16]
Butane	NanoSpace	1	TW-1A	3.5	2015	[17]
R236fa	University of Texas at Austin	110	BEVO-2	3.5	2016	[18]
R236fa	VACCO	10	NanoACE	5.2	2017	[19]
Butane	NanoSpace	1	GomX-4B	8	2018	[20]
R236fa	VACCO	25	MarCO	13.5	2018	[19]
R236fa	University of Texas at Austin	110	ARMADILLO	4	2019	[18]

2.4.1 SNAP-1 (SSTL)

The first flight record of a cold gas microthruster system can be traced back to as early as 2000. Surrey Satellite Technology Ltd. (SSTL) has initiated its first nano-satellite, SNAP-1, in 1999 with the mission as an inspection satellite to capture the image of another satellite, Tsinghua-1. A single-axis propulsion system, placed as close to the satellite's center of gravity as possible, was required to propel the SNAP-1 into a close range of distance to the target satellite.

The SNAP-1 satellite bus consists of three sets of electronic modules, connected in a triangular configuration. The payload panel is placed on top of the modules, which limits the implementation of the typical design configuration of a single central propellant tank. As a result, a unique design, which removed the entire propellant tank and replaced it with a 1.1 m of coiled titanium tube, has been adapted, as shown in Fig. 2.2 [21]. This innovative pipework assembly of 65 cm^3 volume is possible due to the selection of butane as a propellant.

Five possible propellants were considered. While nitrogen and Xenon gas provide the most inferior performance, the need for a heavy high pressurized tank has ruled them out, considering the total mass of the satellite is merely 6.5 kg. Of

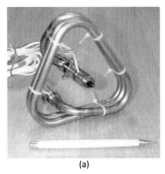 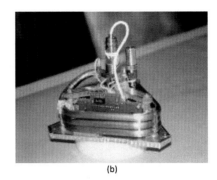

(a)   (b)

FIGURE 2.2

(A) Pipework assembly and (B) integrated cold gas propulsion module of SNAP-1.

Reprint with permission from Chapter 8 - Micropropulsion, in: Z. You (Ed.), Space Microsystems and Micro/Nano Satellites, 2018, Butterworth-Heinemann. pp. 295–339.

the three remaining options, ammonia outperforms propane and butane in terms of total impulse. Safety concerns arose as ammonia is a toxic substance, which requires special handling and hence additional cost. Eventually, butane was chosen due to its lower vapor pressure of 3.8 bar (as opposed to 14.5 bar for propane) at the operating temperature of 40°C. The low pressure provides a good safety margin of 200 for titanium tube and 12 for isolation valve, respectively.

The pipework assembly is connected to a fill valve on one end for filling of butane propellant while a thruster valve on the other end for operational control. An isolation valve is placed in between them to protect against a thruster leakage. The integrated propulsion module, with a total mass of 450 g including 32.6 g of butane, is shown in Fig. 2.2B. The system has successfully been tested in space and raised the orbit of SNAP-1 by 2.6 km. The specific impulse achieved is evaluated as 43 s, which is lower than the theoretical value of 70 s, due to incomplete vaporization of butane that led to the expulsion of liquid phase propellant at the early stage of operation. Nevertheless, the targeted total ΔV of 2 m/s was achieved.

2.4.2 MEPSI (The Aerospace Corporation)

Microelectromechanical system (MEMS) PICOSAT Inspector (MEPSI) mission consists of a tethered pair of picosatellites. After ejection from the Space Shuttle Discovery during the STS-116 mission, the two picosatellites were separated by a tether of 15 feet, which keeps them floating apart. One picosat (Inspector) uses its cold gas propulsion system, developed by The Aerospace Corporation, to get closer to the other picosat (Target) for inspection and download the imagery to the earth ground station.

In any cold gas propulsion system, a pipeline connects one component to the other, for example, propellant tank to valve. The joints, where the pipeline interfaces with the component, are the points for potential leakage if not properly welded. The issue is particularly alarming as welding could be challenging given the very small and limited volume available in picosatellite. The Aerospace Corporation team used stereolithography (SLA) additive manufacturing technology to produce a leak-tight manifold, which includes tanks, plumbing, and nozzle, in one piece, eliminating the distinct joints except the valve connection, as shown in Fig. 2.3.

The fully integrated cold gas propulsion system has five (5) thrusters rated at 20 mN each, uses xenon gas as a propellant, $1 \times 3.6 \times 3.6$ in. in dimension, and weighs 188 g. The system fabricated using Somos 11120 polymeric material was tested to 1000 psi without breaking, which fulfills the requirement by NASA of 2.5 times the maximum design pressure of 115 psig. Another concern with the use of polymeric material in outer space is the outgassing. The Somos 11120 material was sent for outgassing tests, which reported a relatively large total mass loss (TML) of 2.85% and an acceptable collected volatile condensable mass of 0.01%. Additional procedures to heat the SLA-printed manifold to 60°C in a vacuum for 12 h were carried out to remove the volatile elements and hence lower the TML.

The nozzle A in MEPSI cold gas system was operated on orbit as confirmed by the picosatellite rotation rates measured using the onboard triaxial rate sensor. The impulse of 2.6×10^{-3} Ns was evaluated based on the consumption of 0.2 cc xenon gas discharged at 115 psia and a specific impulse of 30 s. However, the mission was short lived as the picosatellites encountered a memory overflow condition after the operator has put them to sleep over the Christmas holiday in 2006. The flight computer went into an infinite loop and the picosatellites never awoke.

FIGURE 2.3

Leak tight manifolds fabricated using SLA additive manufacturing technology.

Reprint with permission from K. Lemmer, Propulsion for CubeSats, Acta Astronaut. 134 (2017) 231–243.

2.4.3 CanX-2 and CanX-4/5 (UTIAS/SFL)

Canadian Advanced Nanospace eXperiment 2 (CanX-2) is a 3U CubeSat of $10 \times 10 \times 34$ cm with a mass of 3.5 kg, developed by the University of Toronto, Institute for Aerospace Studies, Space Flight Laboratory (UTIAS/SFL). CanX-2 serves as a precursory mission to develop and test several enabling technologies for the upcoming CanX-4/5 formation flying mission.

One of the key technologies demonstrated in the CanX-2 mission is Nanosatellite Propulsion System (NanoPS), as shown in Fig. 2.4 [23]. It is a cold gas microthruster system, using sulfur hexafluoride (SF_6) as a propellant. The system is 500 g in weight and provides a thrust of 50 mN with a specific impulse of 50 s. It is designed mainly for attitude control maneuvers. Thus, the nozzle is positioned such that the thrust produced by NanoPS induces a major-axis spin on the nanosatellite. CanX-2 was launched into space in 2008. After 1 year of operation, it has met and exceeded all mission objectives.

CanX-4/5 is a dual-nanosatellite formation flying demonstration mission. To achieve and maintain a precise satellite formation, precise relative position determination and accurate thrusting are required. Based on the NanoPS heritage on CanX-2, UTIAS/SFL has built Canadian Nanosatellite Advanced Propulsion System (CNAPS) for CanX-4/5, a pair of two identical nanosatellites of 1.5 kg each.

Similar to NanoPS, CNAPS uses SF_6 as a propellant and is able to achieve a specific impulse of 45 s. Loaded with a propellant capacity of 260 mL, CNAPS is capable of a total ΔV of 18 m/s. It is equipped with four (4) independently controlled thrusters located on one face of the nanosatellite bus, with a combined thrust level in the range of 12.5−50 mN, depending on the chamber pressure. As the thrusters are configured to offset from the center of mass, each thruster can be operated independently and used for momentum management in reducing the build-up of unwanted torque.

FIGURE 2.4

Nanosatellite Propulsion System (NanoPS) payload for CanX-2 mission.

Reprint with permission from K. Sarda, et al., Canadian advanced nanospace experiment 2: scientific and technological innovation on a three-kilogram satellite, Acta Astronaut. 59 (1) (2006) 236−245.

The twin nanosatellites were launched in 2014 onboard the Polar Satellite Launch Vehicle, from India. CNAPS has been successfully used to perform a series of precise, controlled, and autonomous formation flights from 1 km range down to 50 m separation between the two nanosatellites.

2.4.4 Delfi-n3xt (TNO, U. Twente, and TU Delft)

The Delfi-n3Xt is a 3U CubeSat, developed under the Delfi program by the Delft University of Technology (TU Delft). One of its mission objectives is to prequalify the $T^3\mu PS$, a cold gas micropropulsion system jointly developed by TNO, TU Delft, University of Twente and SystematIC Design BV. MEMS technology has been used extensively to develop an integrated propulsion system of valve, nozzle, pressure transducer, and filter, as schematically shown in Fig. 2.5.

An innovative way to store the gaseous propellant in a form of solidified grain has been developed. Upon heating the solid grain (Fig. 2.6A) in the Cold Gas Generator (CGG), nitrogen gas is released and flows into a plenum for pressurization. The thrust of up to 6 mN can be produced by opening the valve for the compressed gas to exit from the nozzle. The flight model of $T^3\mu PS$, weights 120 g, is shown in Fig. 2.6B.

The system has been tested in orbit shortly after the launch in 2008. Two out of six CGGs were successfully ignited using 10.1 and 10.3 W of electrical power, respectively, before the failure in the ignition train. A spike in the plenum pressure confirms the release of nitrogen gas from the solid grain. A drop in pressure transducer reading after the valve was opened affirms the thrust generation.

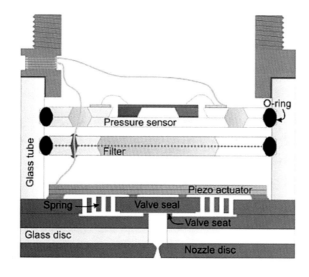

FIGURE 2.5

Schematic of cold gas micropropulsion system in Delfi-n3xt nanosatellite.

Reprint with permission from Delfi Space.

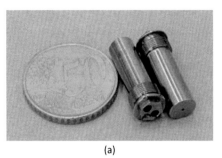

(a)   (b)

FIGURE 2.6

(A) CGG used to produce nitrogen gas; (B) flight model of $T^3\mu PS$.

Reprint with permission from Delfi Space.

2.4.5 POPSAT-HIP1 (microspace)

Propulsion Operation Proof SATellite-High Performance 1 (POPSAT-HIP1) is a 3U CubeSat of 3.3 kg in mass, developed by Microspace Rapid Pte Ltd. It features a cold gas micropropulsion system for attitude control with eight (8) micronozzle modules, as shown in Fig. 2.7A. The supersonic micronozzle structures are mass-produced using a series of silicon MEMS microfabrication techniques, that is, microlithography, deep reaction ion etching, and anodic bonding. After dicing the silicon wafer, an individual micronozzle chip is produced (Fig. 2.7B). Using 5 bar of Argon gas as propellant, each micronozzle chip can produce a nominal thrust of 1 mN.

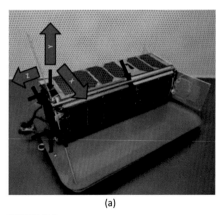

(a)   (b)

FIGURE 2.7

(A) Position of eight micronozzle modules as pointed out by *red arrows*; (B) individual silicon micronozzle chip after dicing.

Reprint with permission from MicroSpace Pte Ltd.

The CubeSat was launched into orbit in 2014. Performance of the cold gas micropropulsion system in angular velocity change, detumble, reduction of angular velocity, and attitude change along two axes has been demonstrated. Maneuvers of 4 degrees/s and up to 10 degrees/s were achieved with the tank pressure of 2 and 5 bar, respectively. During the 9 months in orbit, the system was able to produce a total ΔV up to 3 m/s.

2.4.6 PRISMA, TW-1A and GomX-4B (NanoSpace)

Prototype Research Instruments and Space Mission technology Advancement (PRISMA) is a formation flying and rendezvous technologies demonstration mission, led by Swedish Space Corporation. It consists of two spacecraft. The MAIN spacecraft is highly maneuverable, equipped with three different types of propulsion systems, and hence capable of performing a series of maneuvers around the smaller TARGET spacecraft.

The MAIN spacecraft uses the conventional hydrazine reaction control system, with ΔV capacity of 120 m/s, for most of the maneuvers. Simultaneously, another two propulsion systems, that is, high-performance green propulsion and MEMS-based cold gas micropropulsion, were included as a technology demonstration.

While PRISMA, weighing 145 kg, is not in the nanosatellite category, it has successfully flight demonstrated the functionality and performance of the first fully integrated cold gas micropropulsion system developed by NanoSpace using MEMS technology. The thruster pod consists of a silicon wafer stack with four (4) micromachined microthrusters and integrated with MEMS-based flow control valves, filters, and heaters (Fig. 2.8). Each thruster is capable of producing thrust in the range of 10 μN to 1 mN, using nitrogen gas as a propellant. It is noteworthy that the use of MEMS technology has not only miniaturized the size of the system significantly but also reduced the power consumption of the valve to 3 W per thruster, which is within the power budget of a nanosatellite.

Following the successful technology demonstration in the PRISMA mission, NanoSpace has developed a MEMS-based cold gas micropropulsion system for 3U CubeSat, named as NanoProp CGP3 (Fig. 2.9). The system occupies 0.5 U of volume, uses butane gas as a propellant, and weighs only 350 g (wet mass). There are four (4) individual thrusters, producing 1 mN of thrust each and thrust resolution as low as 10 μN, suitable for both translational maneuvers and attitude control. It has been flight proven and used in formation flying demonstration in TW-1A, a 3U CubeSat developed by Shanghai Engineering Center for Microsatellites, in 2015.

The development of MEMS-based cold gas micropropulsion system for nanosatellites in NanoSpace continues, the first cold gas propulsion system compatible with 6U CubeSat has been developed. The system shares some similarities, such as butane gas as a propellant, four (4) individual thrusters, and 1 mN thrust for each thruster, with NanoProp CGP3 to capitalize on the rich heritage from previous missions. It occupies a volume of $200 \times 100 \times 50$ mm and weighs 900 g (wet mass). The two tanks configuration provides additional redundancy to the system.

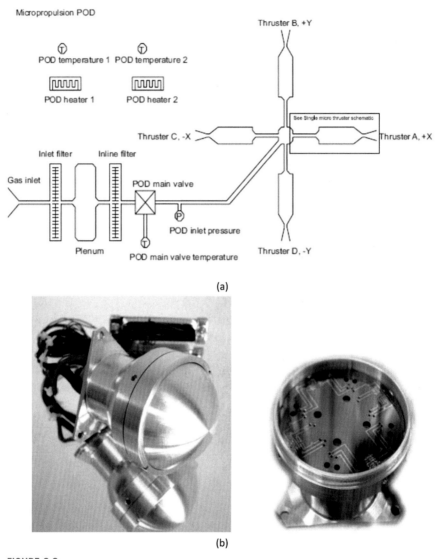

(a)

(b)

FIGURE 2.8

(A) Schematic of MEMS-based micropropulsion system developed by NanoSpace;
(B) thruster pod assembly, 44 mm in diameter and 51 mm in height.

Reprint with permission from T.-A. Grönland, et al., Miniaturization of components and systems for space using MEMS-technology. Acta Astronaut. 61 (1) (2007) 228–233.

The system, named as NanoProp 6U, has been flight proven in the GomX-4 mission in 2018. The mission consists of two 6U CubeSats, GomX-4A and GomX-4B, to demonstrate the key technologies required for large satellite

FIGURE 2.9

NanoProp CGP3 cold gas micropropulsion system developed by NanoSpace for 3U CubeSat [22].

formations in the future. Only GomX-4B is equipped with the propulsion system to perform a critical maneuver, which increases the separation distance between the two nanosatellites to 4500 km, a critical distance set by Earth's curvature, to carry out the intersatellite radio link experiments.

2.4.7 NanoACE and MarCO (VACCO)

VACCO Industries, a company with more than 60 years of experience in engineering and manufacturing of various fluid control components for space applications, has developed a few cold gas micropropulsion systems, specifically for CubeSats. It has supplied its latest Reaction Control Propulsion Module to Tyvak Nanosatellite System Inc for a 3U CubeSat mission, NanoACE. The mission is an internal development program of Tyvak to validate technologies that will be used in future missions.

The module is a self-contained unit and occupies the center of the 3U CubeSat. It has a total of eight (8) thrusters located at the four corners for rendezvous and proximity operations. Each thruster is designed to produce 25 mN of nominal thrust using R236fa as a propellant. The system has been flight proven in 2017.

VACCO has also developed a smart, self-contained micropropulsion system (MiPS) for the first interplanetary CubeSats, Mars Cube One (MarCO). The mission consists of two 6U CubeSats to function as a communication relay for the InSight lander.

The two experimental CubeSats were launched in 2018. After 7 months of cruising together with InSight lander, they survived the trip to deep space. However, the twin CubeSats were last contacted on December 29, 2018. Since then, the engineers have lost communication with the CubeSats. No experiments are conducted for the MiPS system, which has four (4) thrusters with 25 mN nominal thrust, using R236fa as a propellant and a designed total impulse of 775 Ns.

2.4.8 BEVO-2 and ARMADILLO (University of Texas at Austin)

The University of Texas at Austin's Satellite Design Lab has developed a cold gas micropropulsion system for Bevo-2, a 3U CubeSat. The system was manufactured using additive manufacturing technique, SLA, which allows the propellant tank, secondary tanks, internal piping, and nozzle encased in a single block.

The system uses R236fa as a propellant, which is stored in the main tank and flow into the plenums for a complete expansion to gases before exhausting to vacuum space through the nozzle (Fig. 2.10). The thruster is producing thrust in the range of 110–150 mN for a temperature range of 24–85°C. The total mass of the system is 400 g, including 90 g of propellant.

According to the mission planning, the Bevo-2 was scheduled to deploy together with another bigger nanosatellite, AggieSat-4, of 50 kg from ISS in 2016. After the deployment, they will separate in orbit. A proximity operation will be executed thereafter. Unfortunately, the BEVO-2 CubeSat experienced a communication issue and was not activated.

The same cold gas micropropulsion system (Fig. 2.11) was selected in the early stage of another 3U CubeSat mission, ARMADILLO, by the University of Texas at Austin in collaboration with Baylor University.

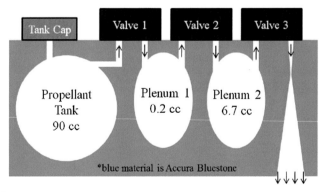

FIGURE 2.10

Schematic of thruster design.

Reproduce with permission from S. Wu, et al., TW-1: a cubesat constellation for space networking experiments.

in: 6th European CubeSat Symposium, 2014.

FIGURE 2.11

Cold gas micropropulsion system featured in Bevo-2 and ARMADILLO CubeSats.

Reproduce with permission from Texas Spacecraft Laboratory.

2.5 Challenges and future

Other than the aforementioned systems that had successfully flight proven, we expect to see more cold gas micropropulsion systems adopted in a wide range of space missions with a scheduled launch in 2019 and 2020. Some of the examples are listed in Table 2.4.

Because of the use of its proprietary Chemically Etched Micro Systems (ChEMS) fabrication technique, cold gas micropropulsion systems from VACCO are generally producing thrust in the range of 10−25 mN. Coupled with its rich heritage and expertise in feed systems, valves, and propellant storage tanks, VACCO's cold gas micropropulsion systems are robust and reliable, particularly suitable for

Table 2.4 Summary of future missions with cold gas micropropulsion system.

Gas	Manufacturer	Thrust (mN)	Mission	Satellite	Expected launch	Reference
R236fa	VACCO	25	CuSP	6U CubeSat	2019	[19]
R236fa	VACCO	25	Omotenashi	6U CubeSat	2019	[19]
R134a/ ADN	VACCO	25	ArgoMoon	6U CubeSat	2019	[19]
R236fa	VACCO	10	CPOD	3U CubeSat	2020	[19]
R236fa	VACCO	25	NEA Scout	6U CubeSat	2020	[19]
R236fa	Georgia Institute of Technology	50	BioSentinel	6U CubeSat	2020	[24]

high-value future missions of 3 and 6U CubeSats. NanoSpace is another company that offers matured and flight-proven cold gas micropropulsion systems. Differ from VACCO's systems, NanoSpace is using MEMS microfabrication techniques, which are able to produce nozzle with dimensions in the order of tenths of micrometers. Thus, the thrust level of NanoSpace's systems can be as low as 1 mN.

While the cold gas micropropulsion systems from both companies are highly integrated and matured, the unit cost might not be affordable for some first-time Cube-Sat missions by universities, which usually have a relatively limited budget. Thus, we have also seen a few cold gas micropropulsion systems developed using 3D additive manufacturing technology, which is very cost effective. Although the material used might not be fully space qualified, it is acceptable for short missions. In addition, 3D additive manufacturing technology is rapidly developing. New manufacturing capabilities and materials will have positive impacts on the future development of cold gas micropropulsion systems based on additive manufacturing.

In the pursuit of a lower thrust level for implementation of cold gas micropropulsions systems in smaller nanosatellites as primary propulsion systems or to provide more precise attitude control capability in bigger nanosatellites, a smaller nozzle (known as micronozzle) is the key enabling component. The subsequent subsections will elaborate on the miniaturization of the nozzle using MEMS technology. Followed by some insights on the optimization of micronozzle design to address the low-performance efficiency in micronozzle.

2.5.1 Miniaturization of nozzle via MEMS approach

With volume in nanosatellites being extremely valuable, it is necessary to reduce the size of propulsion systems as much as possible. With micromachining methods such as micromilling or electrical discharge machining (EDM) methods, one is able to generate features as small as a few hundred microns. However, would one want to further reduce the size of the various propulsion subsystems, alternative manufacturing methods are required. For this further reduction in size, MEMS manufacturing techniques meet the requirements and are generally used in mircopropulsion.

The manufacturing techniques applied for MEMS systems are the same as those used for the semiconductor industry. The smallest attainable feature size is a few nanometers but is rapidly changing, as the semiconductor industry is ever pushing the boundaries to manufacture smaller transistors to keep up with Moore's law. While MEMS manufacturing techniques allow for further reduction of the propulsion system size, they also generate additional limitations. Firstly, the selection of materials is limited as most MEMS manufacturing techniques are designed around the usage of silicon or its derivatives. Although there are some studies that use ceramic as structural materials [25,26], this may not offer an obvious advantage for cold gas micropropulsion systems as they are not operating at high temperatures.

Most micropropulsion systems built with MEMS manufacturing techniques utilize deep reactive ion etching (DRIE). DRIE is a highly anisotropic etching method that is able to generate features with extremely high aspect ratios. DRIE obtains this by alternating between two processes: a short isotropic plasma etch and the deposition of a passivation layer. In the isotropic plasma etch, a plasma, which is made of an appropriate etchant gas, is shot/accelerated perpendicular to the top of the wafer. The ions and free radicals will chemically react on the surface of the wafer and isotropically etch the material away. To generate a useable shape during this etching step a so-called "mask is applied on the top side prior to the DRIE process. This "mask" is a thin deposited layer on top of the wafer with the to be etched shape, e.g., the microthruster, etched out of it. This thin deposited layer is made from a material that will not be etched by the etchant plasma used in the DRIE etch. This way one etches the desired shape into the silicon wafer during the plasma etch step of the DRIE etch. As mentioned this plasma etch step isotropically etches the material away. This would mean that the desired shape would lose all its distinction after a longer etch. As such to generate the necessary depth of the etch while still maintaining a distinctive shape the plasma etch is alternated with a passivation deposition step. This deposition step will deposit an inert material, often some form of polymer, that will prevent any further chemical etching. This adheres to the sidewalls and the bottom of the etched trench in the previous plasma etch step. In the plasma etch the accelerated ions blast away the polymer from the bottom of the trench opening the silicon substrate to the chemical plasma etch. As such one will continue to etch toward the bottom while maintaining the desired profile. This repeated usage of an isotropic plasma etch does result in so called "scallops" on the sidewalls (Fig. 2.12) [27].

Additionally, the manufacturing techniques etch a prescribed pattern into a silicon wafer in the direction perpendicular to the plane on which the pattern is described. Therefore, one is unable to generate a pattern in the etch direction.

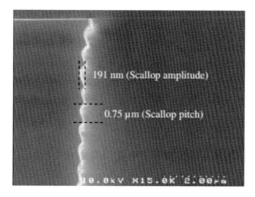

FIGURE 2.12

SEM image of DRIE etched sidewall.

Reproduce with permission from A.M. Malik, et al., Deep reactive ion etching of silicon moulds for the fabrication of diamond x-ray focusing lenses, J. Micromech. Microeng. 23 (12) (2013) 125018.

This results in micronozzles that cannot be axisymmetric: all micronozzles manufactured using MEMS techniques have a rectangular cross-section, as shown in Fig. 2.13 [28].

Various alternative MEMS manufacturing methods such as femtosecond laser machining (FLM) and micro-EDM milling are being researched to be able to manufacture axisymmetric micronozzles. Both methods show promising results to generate axisymmetric nozzles (Fig. 2.14). However, these manufacturing methods are not without drawbacks. The FLM generated geometries are experiencing a significant off-axis thrust component, which was generated due to the fact the axis was not perfectly aligned [29]. The micro-EDM method experienced quite some difficulty to achieve a smooth throat area (Fig. 2.15). Additionally, the throat sizes that one is able to achieve using this method are a multitude larger than that of more conventional methods [30]. Despite these shortcomings, the initial results of these tests show great promise and future developments could result in a shift toward these axisymmetric manufacturing methods. However, due to the wide availability of DRIE equipment, and the well-understood DRIE process, the industry's standard as of now is still DRIE manufactured nozzles with the consequent limitation of producing nozzles having rectangular features.

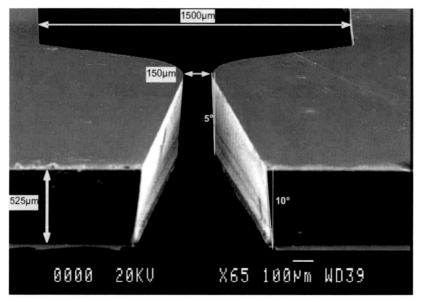

FIGURE 2.13

Rectangular cross-section of micronozzle fabricated using DRIE.

Reproduce with permission from A. Chaalane, et al., A MEMS-based solid propellant microthruster array for space and military applications, J. Phys. Conf. 660 (2015) 012137.

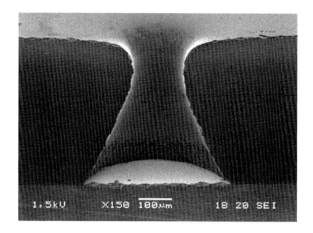

FIGURE 2.14

Axisymmetric 3D micronozzle fabricated using FLM method.

Reproduce with permission from M.C. Louwerse, et al., Nozzle fabrication for micropropulsion of a microsatellite, J. Micromech. Microeng. 19 (4) (2009) 045008.

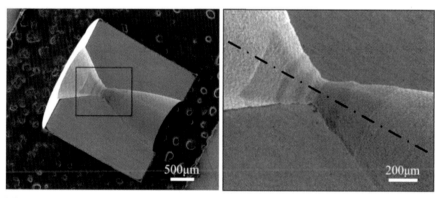

FIGURE 2.15

Axisymmetric 3D micronozzle fabricated using micro-EDM method.

Reproduce with permission from H. Li, et al., Fabrication of ZrB2–SiC–graphite ceramic micro-nozzle by micro-EDM segmented milling, J. Micromech. Microeng. 28 (10) (2018) 105022.

2.5.2 Optimization of micronozzle design

With the usage of MEMS manufacturing techniques, propulsion subsystems were reduced to extremely small sizes with the design nozzle throat being as narrow as 10 μm. With these tiny dimensions, the throat Reynolds number, which is defined by Eq. (2.5), will typically be below 10,000. This is various orders of magnitude lower than the Reynolds number encountered in conventional macro nozzles where throat Reynolds numbers exceed one million.

$$Re_t = \frac{\rho_t V_t D_t}{\mu_t} \tag{2.5}$$

Past research, both numerical and experimental, showed a staggering decrease in nozzle efficiency with decreasing throat Reynolds numbers [31,32]. Especially once the throat Reynolds numbers start to drop below 1000, the decrease in efficiency starts to accelerate, attributed to viscous losses where the thickness of the boundary layer is of importance. In conventional macro nozzles, the thickness of the boundary layer is negligible compared to the overall size of the nozzle. However, once the overall size of the micronozzle starts to decrease, the size of the boundary layer is no longer negligible. In certain cases, the boundary layer can grow to such proportions that it starts to occupy the entire nozzle cross-section (Fig. 2.16), leading to a fully subsonic flow at the divergent section of the nozzle [33]. This boundary layer constricts the effective useable nozzle cross-section and is therefore detrimental to the overall nozzle efficiency. This effect is only amplified at lower throat Reynolds numbers where the boundary layer grows in size.

Past efforts focused on mitigating these losses by using alternate geometries in the nozzle divergent [34,35]. Changing the design parameters, such as the divergence half angle and nozzle length [31,33], of conventional nozzle geometries, one is able to mitigate a significant amount of the viscous losses. However, an optimal geometry exists for each situation and ideally one would like to find this optimal design depending on the initial requirements and constraints. As manufacturing and experimentally testing the range of possible variations of nozzles would be very expensive, tedious, and difficult to perform accurately, numerical simulations are very valuable in obtaining a first-order determination of the optimum design.

However, numerical simulations are not without complications. Due to the extremely small characteristic sizes and low pressures (near-vacuum space environment at the nozzle exit) encountered in micronozzles the continuum assumption, on which most common CFD simulation tools are based, can start to break down, limiting the usability of a lot of common simulation tools. The degree to which

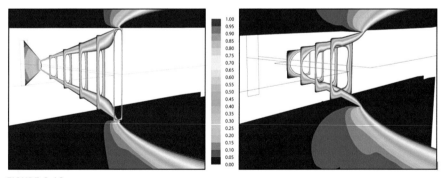

FIGURE 2.16

For $Re \sim 160$ the subsonic layers merge and the entire cross-section of the flow is subsonic. For $Re \sim 320$, only a very narrow region in the center of the flow remains supersonic.

Reproduce with permission from K.H. Cheah, J.K. Chin, Performance improvement on MEMS micropropulsion system through a novel two-depth micronozzle design, Acta Astronaut. 69 (1) (2011) 59–70.

the continuum assumption breaks down can be indicated by introducing the Knudsen number. The Knudsen number is defined by

$$Kn = \frac{\lambda}{L} = \sqrt{\frac{\gamma \pi}{2}} \frac{M}{Re} \qquad (2.6)$$

where λ is the mean free path of a molecule, L is the characteristic length of the flow, γ is the specific heat ratio, M is the local Mach number, and Re is the local Reynolds number.

This number indicates the amount of rarefaction of a flow, a low Knudsen number ($Kn < 0.01$) is indicative of a flow where the continuum assumption is still valid [36]. However, a flow with a Knudsen number above 0.01 will start to show some rarefaction effects. This Knudsen number is inversely related to the Reynolds number, and thus higher Knudsen numbers can be expected in micronozzle flows due to the low Reynolds number.

2.5.2.1 Past developments

With the continuum assumption possibly breaking down, simulations based on the Navier–Stokes equations will start to produce less accurate results. Thus, it was proposed to simulate the flow in the micronozzle using Direct Simulation Monte Carlo (DSMC) [37]. This method is a discrete probabilistic method that simulates single particles in a volume of fluid, each representing a predetermined amount of physical molecules. Through this approach, one is able to simulate a flow for the full range of Knudsen numbers, as the simulated flow is no longer a continuous medium but rather a collection of particles. However, the computational cost of this method is orders of magnitude higher than methods utilizing the continuum assumption.

Micronozzle optimization studies require many simulations to investigate the parametric influence of a selected design parameter on the nozzle performance. This, combined with the fact that computational resources were very limited in the preinformation age era, led to most optimization studies being done using Navier–Stokes-based solvers. Additionally, these optimization studies mostly focused on 2D simulations to limit the computational cost. The validity of these simulations was still kept by utilizing nozzle backpressure that was much higher than vacuum, that is, 1000 Pa. By utilizing this higher backpressure, one ensures that the Navier–Stokes-based solver remains in a Knudsen number regime where it produces valid results [36].

This Knudsen validity regime threshold can be accurately determined through DSMC simulations and is currently a topic of hot debate in the literature. Still, by means of these former 2D Navier–Stokes-based analysis, the following important design considerations could be found:

- Angles of nozzle half divergence larger than the conventional 15 degrees deliver better-performing micronozzles at lower Reynolds numbers. For throat Reynolds numbers below 1000, half angles of 30 degrees and 45 degrees become the most commonly used, with higher angles performing better at lower Reynolds numbers [31,32,36].
- Bell nozzle profiles do not provide higher efficiencies compared to linear nozzles and therefore are not worthwhile to investigate further [34].

2.5.2.2 Current developments

In the last decade, the computational resources available to academic researchers have dramatically increased. As such, 2D numerical simulations using DSMC became more common. Minor micronozzle studies were performed using DSMC; however, large-scale optimization studies still remained too computational costly.

In addition, various efforts were made to make DSMC more computationally efficient for micronozzle simulations through the usage of a hybrid Navier-Stokes-DSMC solver. This solver would simulate the high-density regions, which are very computationally costly if performed with DSMC, using Navier–Stokes. Downstream in the micronozzle, when the pressure drops and thus the Knudsen number rises, the solver switches to DSMC. By utilizing this approach, one is able to achieve the same accuracy of DSMC with as little as 25% of the computational cost compared to a full DSMC simulation [38]. This hybrid approach has great potential for future efforts but, as of today, no optimization or design efforts are run with such a hybrid solver as most research using these solvers is on confirming their accuracy such that they can be used for future research.

However, the preference often still lies with the Navier–Stokes-based solvers for their reasonable accuracy and low computational cost. 3D numerical simulations using Navier–Stokes solvers started to become more common. These simulations found that the 3D micronozzles would significantly underperform their 2D counterpart. This was explained due to the additional losses caused by the presence of a boundary layer that grows on the end walls which previously, in 2D simulations, were neglected.

The influence of the most common design parameters, such as nozzle depth, divergence half angle, and nozzle length, started to be covered extensively given the more powerful simulation means available. Therefore, current developments start to look at novel alternative geometries to further increase the micronozzle efficiency. Examples of such alternative geometries are double-depth linear nozzles, the linear microaerospike nozzle, and double depth linear aerospike nozzles.

Double-depth linear nozzles have a sudden increase in the nozzle depth immediately after the throat. This increase in depth aims to make the boundary layer that grows on the end wall in the divergent section of the nozzle relatively of lesser importance, as it occupies less of the nozzle cross-section. Through this method, performance improvements of up to 5% were realized [33].

The aerospike design is interesting for micronozzles, not much for its ambient pressure compensation features, but more for the fact that the divergent section is an external flow body. This not only means that the flow is able to freely expand to the ambient pressure but also that the flow is not bounded in the etch direction by end walls. Therefore, due to the absence of these walls, the nozzles will also not have any related losses, which is expected to result in higher nozzle efficiencies. 2D numerical studies [35,39] showed great promise for this nozzle geometry, with it generally outperforming the linear nozzles.

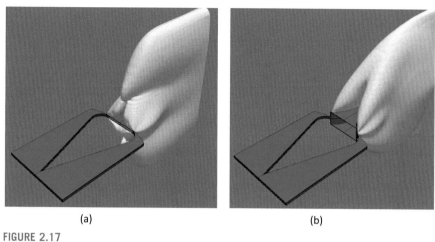

(a)   (b)

FIGURE 2.17

(A) Single depth aerospike nozzle: curling of flow expansion near the edge for $Re_t \approx 2861$. This yields excessive frictional loss. (B) Double-depth aerospike nozzle: Flow expansion over the contour for $Re_t \approx 2861$.

However, 3D numerical studies [39,40] showed that excessive losses over the edge, visible in Fig. 2.17A, cause three-dimensional aerospike designs to severely underperform compared to their two-dimensional counterparts, and in some cases even perform worse than linear nozzles.

A solution for this issue can be represented by a double-depth aerospike nozzle [40]. By increasing the depth of the center body of the aerospike, one is able to force the exiting flow to expand over a contour in contrast to curling over the edge of the center body (Fig. 2.17B). This design solution has shown great potential in 3D numerical simulations, with efficiency improvements of over 20% compared to linear nozzles (Fig. 2.18).

With the emergence of these new alternative micronozzle geometries of great potential in terms of performance and efficiency, the future of micronozzles looks bright. Furthermore, it is expected that further optimization studies on these new geometries could bring even further improvements in nozzle efficiency. Additionally, future computational advancements could make hybrid-DSMC optimization studies feasible, which would allow numerical studies to be performed with near-vacuum ambient back-pressures and, thus, closer to the actual operating conditions of these nozzles in space.

Additionally, the manufacturing techniques etch a prescribed pattern into a silicon wafer in the direction perpendicular to the plane on which the pattern is described. Therefore, one is unable to generate a pattern in the etch direction. This results in micronozzles that cannot be axisymmetric: all micronozzles manufactured using MEMS techniques have a rectangular cross-section.

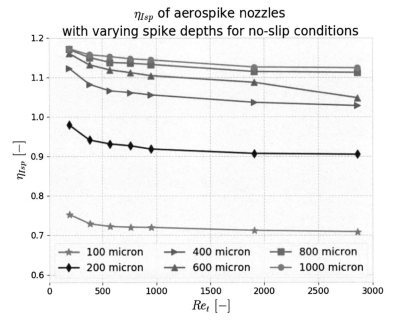

FIGURE 2.18

Specific impulse efficiency of the double depth aerospike nozzles.

References

[1] Sutton, G.P., Biblarz, O., 2001. Rocket Propulsion Elements, seventh ed. John Wiley & Sons.

[2] Mueller, J., et al., 2010. Survey of propulsion options for cubesats. In: 57th JANNAF Propulsion Meeting.

[3] Coates, A.J., et al., 2003. In-orbit results from the SNAP-1 nanosatellite and its future potential. Philos. Trans. R. Soc. London, Ser. A 361 (1802), 199–203.

[4] Nguyen, H., Kohler, J., Stenmark, L., 2002. The merits of cold gas micropropulsion in state-of-the-art space missions. In: 53rd International Astronautical Congress.

[5] Hejmanowski, N., et al., 2015. CubeSat high impulse propulsion system (CHIPS). In: 62nd JANNAF Propulsion Meeting.

[6] Hejmanowski, N., et al., 2016. CubeSat high impulse propulsion syste (CHIPS) design and performance. In: 63rd JANNAF Propulsion Meeting.

[7] Gibbon, D., Ward, J., Kay, N., 2000. The design, development and testing of a propulsion system for the SNAP-1 nanosatellite. In: 14th AIAA/USU Conference on Small Satellites.

[8] Gibbon, D., Underwood, C.I., 2001. Low cost butane propulsion systems for small spacecraft. In: 15th AIAA/USU Conference on Small Satellites.

[9] Hinkley, D.A., 2008. Novel cold gas propulsion system for nanosatellites and picosatellites. In: 22nd AIAA/USU Conference on Small Satellites.

[10] Rankin, D., et al., 2005. The CanX-2 nanosatellite: expanding the science abilities of nanosatellites. Acta Astronaut. 57 (2), 167—174.

[11] Sarda, K., et al., 2008. Canadian advanced nanospace experiment 2 - on-orbit experiences with a 3 kg satellite. In: 22nd AIAA/USU Conference on Small Satellites.

[12] Grönland, T.-A., et al., 2007. Miniaturization of components and systems for space using MEMS-technology. Acta Astronaut. 61 (1), 228—233.

[13] Persson, S., Veldman, S., Bodin, P., 2009. PRISMA—a formation flying project in implementation phase. Acta Astronaut. 65 (9), 1360—1374.

[14] Guo, J., Bouwmeester, J., Gill, E., 2016. In-orbit results of Delfi-n3Xt: lessons learned and move forward. Acta Astronaut. 121, 39—50.

[15] Bonin, G., et al., 2015. CanX-4 and CanX-5 precision formation flight: mission accomplished. In: 29th AIAA/USU Conference on Small Satellites.

[16] Manzoni, G., Brama, Y.L., 2015. Cubesat micropropulsion characterization in low earth orbit. In: 29th AIAA/USU Conference on Small Satellites.

[17] Wu, S., et al., 2014. TW-1: a cubesat constellation for space networking experiments. In: 6th European CubeSat Symposium.

[18] Arestie, S., Lightsey, E.G., Hudson, B., 2012. Development of A Modular, cold gas propulsion system for small satellite applications. Journal of Small Satellites 1 (2), 63—74.

[19] Cardin, J., Day, C., 2020. A standard micro propulsion system for CubeSats. In: CubeSat Developers Workshop.

[20] Perez, L.L., Koch, P., Walker, R., 2018. GOMX-4 — the twin European mission for IOD purposes. In: 32nd AIAA/USU Conference on Small Satellites.

[21] Chapter 8 - Micropropulsion. In: You, Z. (Ed.), 2018. Space Microsystems and Micro/Nano Satellites. Butterworth-Heinemann, pp. 295—339.

[22] Lemmer, K., 2017. Propulsion for CubeSats. Acta Astronaut. 134, 231—243.

[23] Sarda, K., et al., 2006. Canadian advanced nanospace experiment 2: scientific and technological innovation on a three-kilogram satellite. Acta Astronaut. 59 (1), 236—245.

[24] Stevenson, T., Lightsey, G., 2016. Design and characterization of a 3D-printed attitude control thruster for an interplanetary 6U cubesat. In: 30th AIAA/USU Conference on Small Satellites.

[25] Ranjan, R., et al., 2018. Cold gas propulsion microthruster for feed gas utilization in micro satellites. Appl. Energy 220, 921—933.

[26] Cheah, K.H., Chin, J.K., 2014. Fabrication of embedded microstructures via lamination of thick gel-casted ceramic layers. Int. J. Appl. Ceram. Technol. 11 (2), 384—393.

[27] Malik, A.M., et al., 2013. Deep reactive ion etching of silicon moulds for the fabrication of diamond x-ray focusing lenses. J. Micromech. Microeng. 23 (12), 125018.

[28] Chaalane, A., et al., 2015. A MEMS-based solid propellant microthruster array for space and military applications. J. Phys. Conf. 660, 012137.

[29] Louwerse, M.C., et al., 2009. Nozzle fabrication for micropropulsion of a microsatellite. J. Micromech. Microeng. 19 (4), 045008.

[30] Li, H., et al., 2018. Fabrication of ZrB_2—SiC—graphite ceramic micro-nozzle by micro-EDM segmented milling. J. Micromech. Microeng. 28 (10), 105022.

[31] Louisos, W.F., Hitt, D.L., 2008. Viscous effects on performance of two-dimensional supersonic linear micronozzles. J. Spacecraft Rockets 45 (4), 706—715.

[32] Ketsdever, A.D., et al., 2005. Experimental and numerical determination of micropropulsion device efficiencies at low Reynolds numbers. AIAA J. 43 (3), 633—641.

[33] Cheah, K.H., Chin, J.K., 2011. Performance improvement on MEMS micropropulsion system through a novel two-depth micronozzle design. Acta Astronaut. 69 (1), 59—70.

[34] Louisos, W., Hitt, D., 2008. Numerical studies of thrust production in 2-D supersonic Bell micronozzles. In: 44th AIAA/ASME/SAE/ASEE Joint Propulsion Conference & Exhibit.

[35] Zilic, A., Hitt, D., Alexeenko, A., 2007. Numerical simulations of supersonic flow in a linear aerospike micro nozzle. In: 37th AIAA Fluid Dynamics Conference and Exhibit.

[36] Louisos, W.F., et al., 2008. Design considerations for supersonic micronozzles. Int. J. Manuf. Res. 3 (1).

[37] Stein, W., Alexeenko, A., 2008. Application of the DSMC method for design of a co-axial microthruster nozzle. In: 44th AIAA/ASME/SAE/ASEE Joint Propulsion Conference & Exhibit.

[38] La Torre, F., et al., 2011. Hybrid simulations of rarefied supersonic gas flows in micro-nozzles. Comput. Fluid 49 (1), 312–322.

[39] Pearl, J., Louisos, W.F., Hitt, D.L., 2015. Three-dimensional numerical study of linear plug micronozzles. In: 53rd AIAA Aerospace Sciences Meeting.

[40] Ganani, C.S., 2019. Micronozzle Performance A Numerical and Experimental Study. Delft University of Technology.

Solid-propellant microthruster

3

Akira Kakami, PhD

Associate Professor, Department of Aeronautics and Astronautics, Tokyo Metropolitan University, Hino, Tokyo, Japan

3.1 Introduction

Solid-propellant rockets, which have simpler structures than liquid and hybrid rockets, have been used as launching vehicles and apogee kick motors that are applied to orbit insertion. In 1958, one year after the World's first satellite Sputnik flew in the Earth orbit, the United States of America launched the Pioneer I spacecraft, which had two kinds of solid-propellant engines for the orbit insertion. One is referred to as "Eight Rockets," which were IXS-50 attitude control engines for the Vanguard rocket [1,2]. The solid rocket engine produced a thrust of 222 N with 1-s burning duration and used resin/ammonium perchlorate (AP), Mg/polyisobutadiene/KNO_3 as a fuel, oxidizer, and ignitor. The solid-propellant rockets were installed so that the Pioneer I would be inserted into the Luna orbit [3].

Moreover, in 1964, the Symcom III satellite, the first geostationary satellite, had Thiokol's Starfinder solid-propellant apogee rocket engine (thrust: 5000 N, specific impulse I_{sp}: 274 s) as well as hydrogen peroxide thrusters for the position control.

We can find that solid-propellant engines were installed in spacecraft at the dawn of the space development era. However, today, almost all the powered-flight spacecraft have no solid-propellant thruster but have hydrazine monopropellant thrusters or hydrazine/nitrogen tetroxide bipropellant thrusters. In some spacecraft, electric propulsion, such as Hall thrusters, is applied to gain higher ΔV than monopropellant/bipropellant thrusters. This is because the solid-propellant thrusters have difficulty in the termination and restart of thrust production. Hence, liquid-propellant thrusters and electric propulsion are applied to the modern spacecraft owing to the throttleability, restart capability and specific impulse.

Today, startup companies and universities design microsatellites, and some of the microsatellites or small satellites have small or microthrusters. The solid-propellant thrusters started to attract attention because of the features: simplicity and higher specific impulse than cold gas jets. Accordingly, many groups develop various solid-propellant propulsion. Hence, this section addresses solid propellants and the principles and prototypes of solid-propellant thrusters.

Fig. 3.1 shows a schematic of typical solid-propellant rockets. The conventional solid-propellant rockets are generally comprised of thrust chambers, de Laval

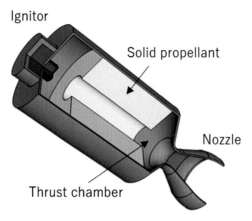

FIGURE 3.1

Solid-propellant rockets.

nozzles, and ignitors. The solid propellant, referred to as a grain, was formed in the thrust chamber. There are various grain geometries, such as end burning, internal burning, and star grains. Thrust profiles are preprogrammable by the grain geometry, while solid propellant is not throttleable (Fig. 3.2). The star and end burning grains show the neutral thrust profile, in which the thrust is almost constant during firing because the burning surface area remains constant. In contrast, the thrust profile is progressive, which means that the thrust is gradually increased during firing for internal burning grain because the port diameter is expanded by combustion, and the burning surface area is increased. As will be mentioned later, most of the solid-propellant microthrusters mainly use end burning type, which can simplify thruster design.

The micropropulsion prototypes have diversity compared with the conventional rocket engines for launching/sounding rockets. Regarding ignition, conventional solid-propellant rocket engines use pyrotechnic ignitors. The ignitor uses pyrolants, which are chemical reactants that readily induce exothermic reaction and produces intense heat but yield low specific impulses when they are applied as a propellant. The pyrolant reaction is generally started with resistive heaters. On the other

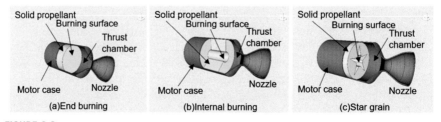

FIGURE 3.2

Solid-propellant grain.

hand, microthrusters use not only resistive heaters but laser diodes as well as pyrotechnic ignitors.

Various propellants have been examined in micropropulsion. As will be shown in the next chapter, the conventional solid-propellant rocket engines use composite solid propellants, which are usually a mixture of butadiene rubber and particulate oxidizer to produce relatively high specific impulse. For some prototypes, explosives and pyrolants were utilized as a propellant to ensure ignition. With regard to fabrication, the micro-electro-mechanical systems (MEMS) technology is also applied to make precise and small thrusters so that we can see the prototypes whose appearance is the integrated circuit chips. Then, a variety of solid propellant are used in micropropulsion, and the prototypes will be addressed in Section 3.5.

3.2 Solid propellants

Table 3.1 presents typical reactants used in solid propellants. Solid propellants are categorized into two types: double-base (DB) propellant and composite propellant; from the microscopic viewpoints, DB propellants have homogeneous structures, whereas composite solid propellants have heterogeneous structure. The main ingredients of DB propellant are nitrocellulose (NC, solid at room temperature) and nitroglycerine (NG, liquid at room temperature). On the other hand, the composite solid propellant contains a fuel binder and particulate oxidizer. Metal powder fuels and particulate combustion catalysts are sometimes added to the propellant in order to adjust burning rates or enhance performances.

Microthrusters use not only these conventional solid-propellant reactants but chemicals that have not been applied. This seems to be due to the acceptance of the short pulse firing and the necessity of rapid and certain ignition under the vacuum environment. The launching and sounding solid rockets need to produce the kN class continuous thrust for seconds and minutes order. Even apogee kick motors should produce thrust for a few tens of seconds; the Star 27 apogee kick motor has 34-s burning time. Hence, the solid propellant for the conventional solid rockets should burn stably without oscillating or diverging thrust chamber pressure, and the stability necessitates the pressure exponent of $n < 1$, as will be mentioned in Section 3.3.1. In contrast, some space propulsion devices produce millisecond impulsive thrust, and accordingly, solid chemicals with $n \geq 1$ become an option either. Moreover, the design where the solid propellant is exposed to the vacuum could have to be used because sealing the nozzle exit could complicate the thruster. However, the solid propellants generally have difficulty in the ignition and could have unacceptable ignition delay under the low backpressure. For such thrusters, readily ignitable chemicals, such as pyrolant, even if $n \geq 1$, would have advantages in micropropulsion. Hence, unconventional propellants have been examined in microthrusters.

Nanothermite is a good example and uses nanoparticles for thermite reaction. The thermite reaction is the chemical reaction in which intense heat is produced using metals and oxides and is used in industries. For example, Al/Fe_2O_3 reaction has been used in welding rails in the construction of railroads. The nanothermite reaction

Table 3.1 Reactants for solid propellants.

Reactants	Chemical formula	Molecular mass	Density	Heat of formation, ΔH_f
Aluminum	Al	29.9815	2700	0
Boron	B	10.81	2080	0
HTPB[a]	$C_{4n}H_{6n+2}O_2$	≈ 2800	900	-0.31 MJ/kg [4]
HTPB binder	$C_{7.075}H_{10.65}O_{0.223}N_{0.063}$	–	–	-0.058 MJ/mol [4]
HTPB binder	$[C_{10}H_{15.4}O_{0.07}]_n (n=1)$	136.5	916	-51.9 kJ/mol [5]
GAP copolymer	$C_{3.5}H_{5.6}O_{1.12}N_{2.63}$	1270	1300	(0.96 MJ/kg[b]) [4]
GAP binder	–	500–5000	1300	957 kJ/kg[b] [6]
GAP binder	$[C_3H_5N_3O]_n (n=1)$	99	1300	117.2 kJ/mol[b] [5]
GAP diol	–	2097	1300	0.28 kcal/g[b] [7]
AP	NH_4ClO_4	117.49	1950	-2.52 MJ/kg [4]
AN	NH_4NO_3	80.043	1725	-4.56 MJ/kg [4]
Potassium nitrate	KNO_3	101.1032	2109	-494.6 kJ/mol [4]
ADN	$NH_4N(NO_2)_2$	124.06	1810	-134.6 kJ/mol [8]
HAN	$[NH_3OH]^+[NO_3]^-$	96.04	1830	-79.68 kcal/mol [9]
NG	$(ONO_2)_3(CH_2)_2CH$	227.09	1590	-1.70 MJ/kg [4]
NC	$C_{18}H_{21}N_{11}O_{38}$ [10]	999.4 [10]	1670	-2.60 MJ/kg [4]
NC(12.6% N)	–	–	–	-2598 kJ/kg [11]
LTNR	$C_6H(NO_2)_3O_2Pb$	450.95	3020 (crystal), 1400–1600 (bulk)	855 kJ/mol for monohydrate [12]
RDX	$C_3H_6N_6O_6$	222.1163	1820 [4]	79.1 ± 5.0 kJ/mol [13]
HMX	$C_4H_8N_8O_8$	296.1551	1900 [4]	$103.\pm 2.8$ kJ/mol [13]

[a] The data are for uncured HTPB. Polymerization generally changes the properties.
[b] Heat of formation is that of GAP prepolymer.

is preferable to microthruster since the reaction is readily initiated, although producing lower theoretical specific impulse than DB and composite propellants. This section introduces typical reactants.

3.2.1 Fuel

Glycidyl azide polymer (GAP) is an azide polymer, which has the polymer that has $-N = N^- = N^+$ group. The group $-N_3$ releases numerous heat during dissociation. Accordingly, the fuel has a positive heat of formation (0.96 MJ/kg) [4] and produces intense heat during combustion (150 cal/g) [7]. GAP diol prepolymer (Fig. 3.3) has a chain-like structure with branches ($n \approx 20$) and liquid form at room temperature (Fig. 3.4). Hexamethylene diisocyanate (HMDI) and trimethylolpropane are blended into the prepolymer to form GAP polymer by cross-linking. Moreover, GAP triol (Fig. 3.5) is used because of their performance. The GAP triol is also cross-linked

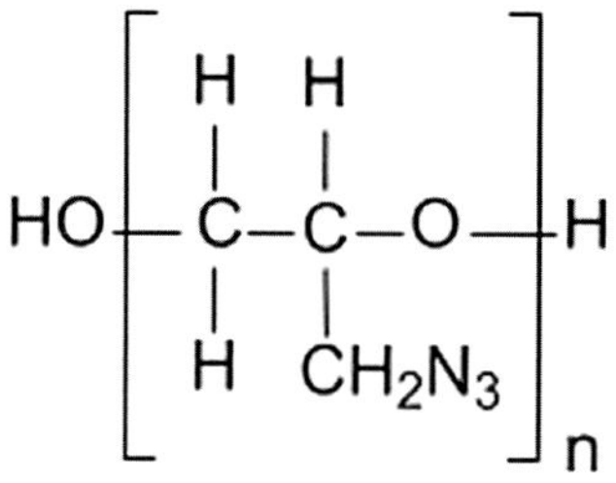

FIGURE 3.3

GAP diol molecule [4].

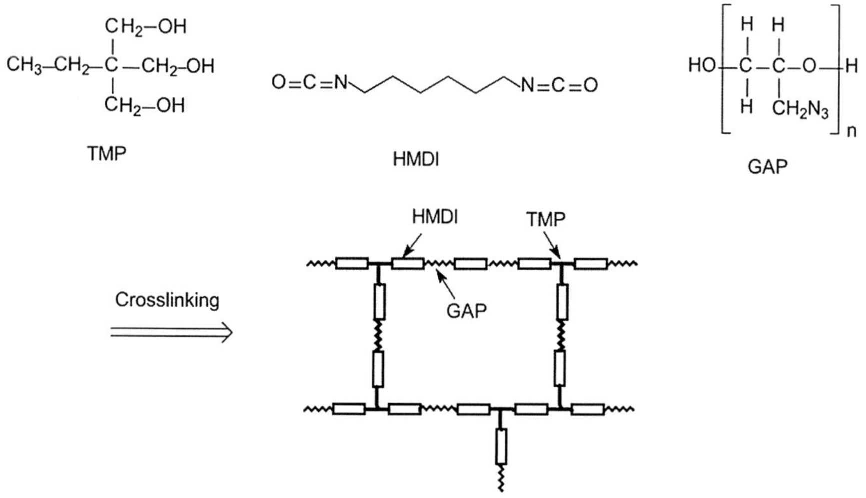

FIGURE 3.4

GAP molecule [4].

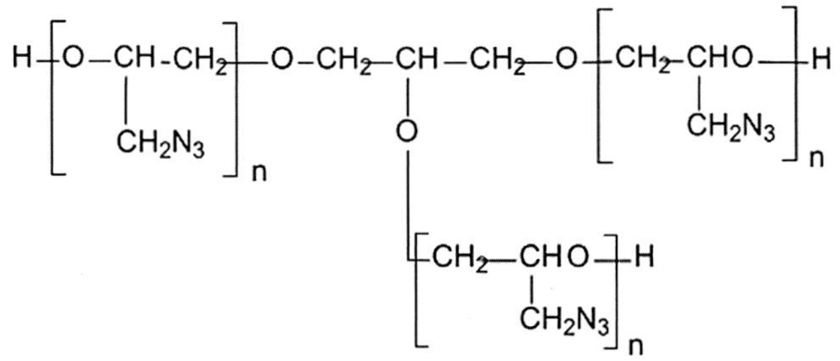

FIGURE 3.5

GAP triol molecule [7].

using HMDI to form the polymer, and the binder molecule shape is more complex than the binder made from the GAP diol prepolymer.

Hydroxyl-terminated polybutadiene (HTPB) is a rubber, which is prepared into polyurethanes owing to elasticity, toughness, and rigidity. Hence, the rubber is made into tires, pannels, and sealants. HTPB has a long chain-like structure, as illustrated in Fig. 3.6, and is a transparent liquid before being cured. Curing agents such as isophorone diisocyanate and dioctyl adipate are added and heated up to 60–70 °C for approximately seven days to form the HTPB solid fuel. Whereas there are various kinds of HTPB polymers that have different lengths of chains, certain types such as R-45M are usually used as HTPB-based propellants.

Besides these polymers, metals are also used as fuels. Aluminum (Al), which is usually used as a general structural material, is high-performance fuel. Al powder is applied to propellants because it has low molecular weight among metals and enormous combustion heat by oxidation. In many cases, Al fuel powders are added into

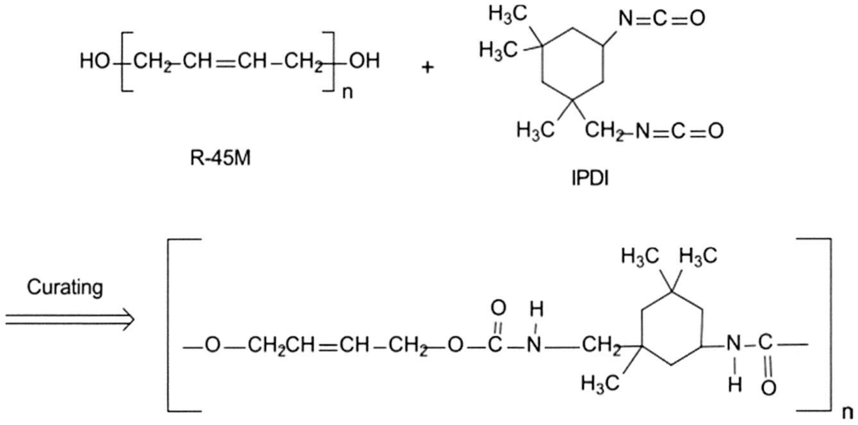

FIGURE 3.6

HTPB molecule [4].

HTPB/AP composite propellant so that the burning rate is increased and thrust is augmented. In many cases, Al powders were added by 20wt%.

Boron (B), which is a nonmetal energetic material and is sometimes categorized into metalloid, has a relatively low molecular weight and large combustion heats through the oxidization process. Though the fuel theoretically produces higher heat of combustion (18.27 kJ/kg) [4], it has difficulty in maintaining stable combustion because the B_2O_3 layer covers the burning surface. Hence, B is not used as rocket propellants but as a ducted-rocket fuel and pyrolant of ignitors.

3.2.2 Oxidizer

AP is a white crystallized salt whose chemical formula is expressed as $[NH4]^+[ClO4]^-$. The crystal is occasionally used in an oxidizer for composite propellants because of high performance and availability, though it is deliquescent and can be degraded owing to the humidity. As will be mentioned later, HTPB/AP propellant is often used for launching/sounding rocket engines. This oxidizer induces an exothermic chemical reaction and emits oxygen, and hence, can be categorized into an oxygen-rich monopropellant. Rapid heating induces AP exothermic decomposition in accordance with the reaction [4]:

$$NH_4ClO_4 \rightarrow NH_3 + HCl + 2O_2 \tag{3.1}$$

Note that Zr and Al are sometimes added into AP-based composite propellants to suppress the combustion instability.

Ammonium nitrate (AN, $[NH4]^+[ClO4]^-$) is a deliquescent white crystallized nitrate salt and has a freezing point of 442 K. In the vicinity of 480 K, an endothermic decomposition is induced as:

$$NH_4NO_3 \rightarrow NH_3 + HNO_3 - 178[kJ/mol] \tag{3.2}$$

In contrast, in the higher temperature region, an exothermic reaction

$$NH_4NO_3 \rightarrow N_2 + 2H_2O + \frac{1}{2}O_2 + 119[KJ/mol] \tag{3.3}$$

is induced to generate oxygen and intense heat [4]. This change in releasing heat complicates AN ignition.

Potassium nitrate (KNO_3) has been used as an ingredient of black powder (BP) from ancient days. The crystal is used as an oxidizer for pyrolant since it has a weak sensitivity to impact and high safety. An exothermic reaction is expressed as:

$$KNO_3 \rightarrow \frac{1}{2}K_2O + NO + \frac{3}{4}O_2 \tag{3.4}$$

Ammonium dinitramide (ADN) is a nitramine, which is a molecule that has $-N-NO_2$ group, is a water-soluble solid oxidizer. The oxidizer yields a little higher theoretical specific impulse and is environment-friendly because it contains no halogen atoms such as chlorine atom. ADN was developed by the USA and USSR independently as a high-performance oxidizer for solid propellants [14], but its solution is used as a liquid propellant by the use of its water-solubility. For example, in LMP-103S monopropellant, methanol is blended into the ADN water solution. Owing to the low toxicity compared with hydrazine, LMP-103S is expected to be a green propellant [15].

Hydroxyl ammonium nitrate (HAN, $[NH_3OH]^+[NO_3]^-$) is a solid ionic oxidizer and deliquescent white crystal [9]. HAN water solution is a nontoxic monopropellant because the HAN water solution has a maximum mixture ratio of 95%. For example, LP-1845 monopropellant, in which HAN is blended with triethyl ammonium nitrate (TEAN), has been used as gun propellant. Although HAN is a potential green monopropellant because of nontoxicity and high concentration ratio, the burning rate is drastically increased above 0.8 MPa. Hence, new HAN-based propellants such as SHP-163 [16] and AF-M315E [17] were developed, such that propellant shows a slower increase in the burning rate at 0.8 MPa.

3.2.3 Other reactants

Besides these fuels and oxidizers, there are reactants that are categorized into explosives. Lead styphnate (LTNR, lead 2,4,6-trinitroresorcinate) has a styphnic acid salt whose chemical formula is $C_6HN_3O_8Pb$. LTNR has a melting point ranging from 260 to 311 °C and has been applied to detonators of fuses for guns.

NG is a nitrate ester, which has the $-O - NO_2$ group, and is used as both propellants and explosives (Fig. 3.7). The reactant is liquid at room temperature, and freezes at 286 K. Desensitizers are usually utilized because NG is sensitive to shocks and induce detonation. NG is generally used with NC to form the DB propellant. Dissociation of $-O - NO_2$ group produces NO_2 gas.

NC is also a nitrate ester and formed by adding nitric acid to cellulose (Fig. 3.8). The characteristics are variable by the number of cellulose chains (n) and the degree of nitration. Though the reactant can be used as a single-base propellant, NC is usually used with NG as the double-base propellant to enhance performance. The enthalpy of formation is dependent on nitration, and expressed using an empirical formula [20,21]:

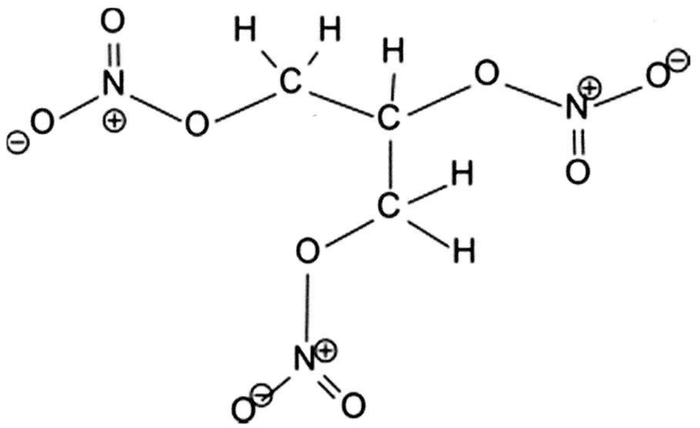

FIGURE 3.7

Nitroglycerin molecule [18].

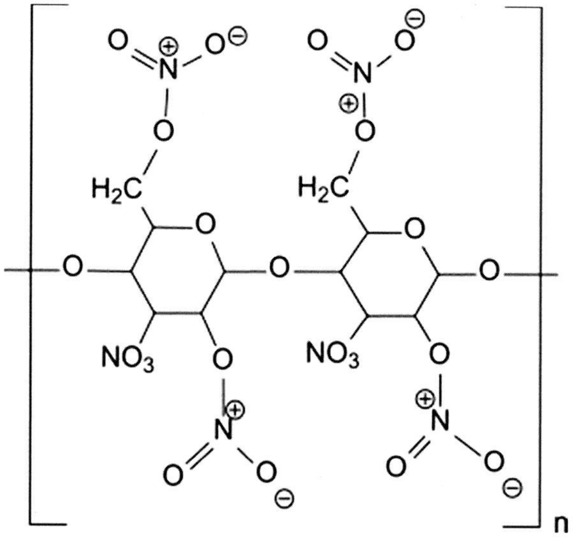

FIGURE 3.8

Nitrocellulose molecule [19].

$$\Delta H_f = -1406.22 + 6231.86f \tag{3.5}$$

The unit is cal/g, and f is the weight-fraction of nitrogen in NC. The molecular structure is $C_6H_{10-x}O_{5+2x}N_x$.

$$x = \frac{162.1430f}{14.0067 - 44.9975f} \tag{3.6}$$

The corresponding molecular weight M is:

$$M = 44.9975 + 162.1430f \tag{3.7}$$

RDX (Research Department eXplosive, cyclonite Fig. 3.9) and HMX (High Melting eXplosive, cyclotetramethylenetetranitramine, Fig. 3.10) are nitramines,

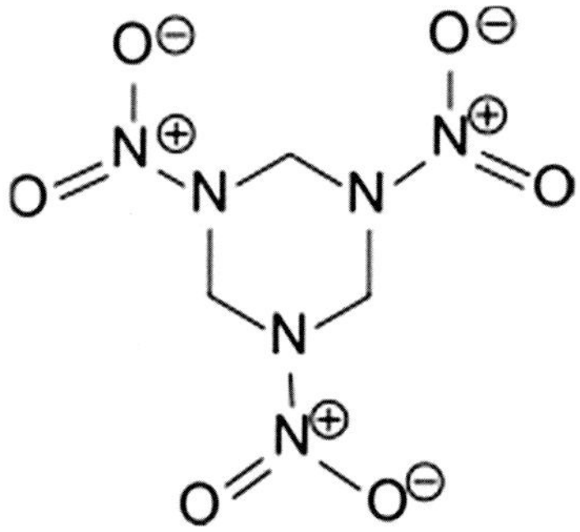

FIGURE 3.9

Research department eXplosive (RDX).

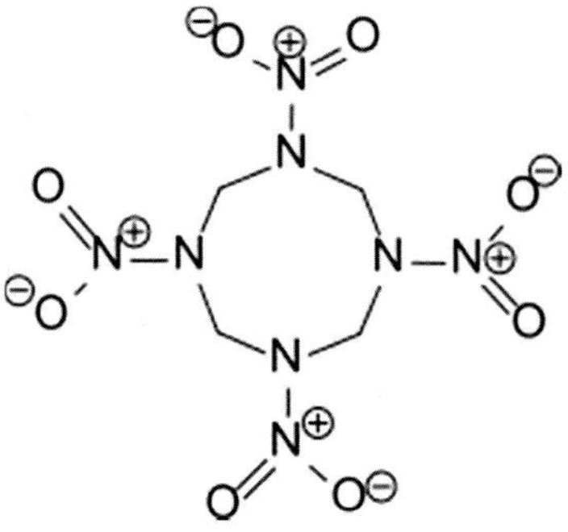

FIGURE 3.10

High melting eXplosive (HMX).

which have $-N-NO_2$ group and are usually categorized into explosives. These reactants yield exothermic reactions.

$$(NNO_2)_3(CH_2)_3 \rightarrow 3CO + 3H_2O + 3N_2 \qquad (3.8)$$

$$(NNO_2)_4(CH_2)_4 \rightarrow 4CO + 4H_2O + 4N_2 \qquad (3.9)$$

As shown in Eqs. (3.8) and (3.9), RDX and HMX produce no surplus oxygen in the dissociation.

3.2.4 Propellants

The HTPB/AP propellants yield a higher specific impulse among composite propellants, and hence, are used as a solid propellant for launching vehicles. Fig. 3.11 shows that theoretical specific impulse and adiabatic flame temperature are dependent on the mixture ratio. Assuming the combustion gas maintains chemical equilibrium at both thrust chamber and nozzle, rocket engines with 1-MPa thrust chamber pressure and 50 expansion ratio theoretically yield the maximum theoretical specific impulse of 340 s for HTPB/AP = 10/90wt%. Moreover, since the diameter of AP particle affects burning rate, three types of AP are mixed: coarse (400−600 μm), medium (50−200 μm), fine (5−15 μm), and ultra-fine (no more than 5 μm).

Other fuels are also added to the HTPB/AP propellant. The addition of metal fuel such as aluminum particle can increase the burning rate and rise theoretical specific impulse so that specific impulse and thrust are spontaneously enhanced. However, completing combustion of Al-based propellants requires sufficient residual time in the thrust chambers, and accordingly, thrust chamber pressure and characteristic length L^* (thrust chamber volume per nozzle throat area) need to be enlarged to enhance the performance. Therefore, the Al fuel could not be appropriate to micropropulsion devices.

Note that the thrusters would produce much lower I_{sp} than the theoretical values shown in the figures, including Fig. 3.11. This is attributable to the non-realistic assumption of chemical equilibrium and the heat loss to the thrust chamber and

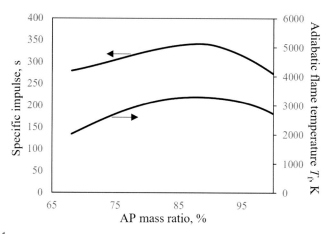

FIGURE 3.11

Theoretical specific impulse and adiabatic flame temperature of HTPB/AP at a thrust chamber pressure of 1MPa and expansion ratio of 50.

nozzle. In the actual thrusters, the combustion gas is accelerated by the nozzles so fast that the chemical reaction is not able to respond to the change in thermodynamic variables such as pressure and temperature, which is induced by the nozzle. In other words, the residual time in the thruster is shorter than the time for the chemical reaction. Accordingly, the combustion gas is expelled from the nozzle before completing the chemical reaction toward the chemical equilibrium, which would have released more chemical enthalpy that augments the kinetic energy of combustion gas.

The heat loss also deteriorates the performance. For achieving equilibrium, the body should be sufficiently heated by the combustion gas. However, the heat is absorbed by the thruster soon after ignition, and the absorbed heat makes no contribution to the thrust production. Hence, decreasing firing period reduces specific impulse, and in nature, the pulse operation types cannot avoid yielding smaller specific impulse.

As shown in Fig. 3.12, potassium nitrate (B/KNO_3) presents an adiabatic flame temperature of 2800 K at the Boron weight ratio of 0.25 and yields a theoretical specific impulse (200 s), which is much lower than that of HTPB/AP propellants. Hence, B/KNO_3 has been used as a pyrolant for pyrotechnic ignitors in conventional solid-propellant rockets. However, it is utilized as a propellant of microthrusters because it is readily ignited in a vacuum. Note that in B/KNO_3 combustion, there are three stages. In the first stage, the B particle reacts with melt KNO_3 to yield KBO_2 and NO with intense heat. Afterward, the remaining potassium nitrate is evaporated to form oxide gas. Finally, KBO_2 is dissolved into K_2O and B_2O_3 [22].

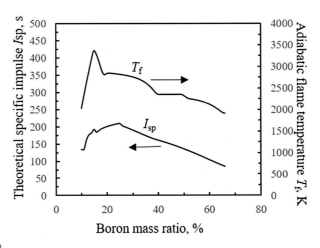

FIGURE 3.12

Theoretical specific impulse and adiabatic flame temperature of B/KNO$_3$ at a thrust chamber pressure of 1 MPa and an expansion ratio of 50.

$$B + KNO_3 \rightarrow KBO_2 + NO \rightarrow \frac{1}{2}K_2O + \frac{1}{2}B_2O_3 + NO \qquad (3.10)$$

The NC/NG propellant is the DB propellant, and the mixture ratio is near the stoichiometric mixture ratio so that the propellant produces only a little smoke. However, the propellant yields a lower theoretical specific impulse than the composite solid propellants. Specific impulse is dependent on the NG mixture ratio, as shown in Fig. 3.13. The NG mixture ratio is usually no more than 50%, whereas

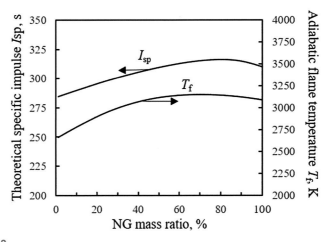

FIGURE 3.13

Theoretical specific impulse and adiabatic flame temperature of NC/NG at a thrust chamber pressure of 1 MPa and an expansion ratio of 50.

the propellant yields the maximum specific impulse in the vicinity of 100% NG concentration. This is because the propellant NG is liquid at room temperature and chemically unstable.

3.3 Solid-propellant propulsion fundamentals

3.3.1 Thrust chamber pressure and stability

The thrust chamber pressure is an essential factor for the design because the excessively high pressure could violate the structure, but higher pressure tends to enhance the specific impulse. Moreover, the time variation in pressure also damages thrusters and has a negative affect on spacecraft and missions. Therefore, the thrust chamber pressure and its stability are necessary for stable thrust production. Then, the chapter discusses the thrust chamber pressure and its stability.

Burning rate, regressing velocity of burning surface, r is empirically expressed as Vieille's law:

$$r = ap^n \tag{3.11}$$

where p is the backpressure, a is a constant that is dependent on initial solid-propellant temperature, and n is a specific value to solid-propellant ingredients and mixture ratio. In general, the burning rate is increased with increasing initial solid-propellant temperature.

Then, assuming that thrust F is produced only by the momentum thrust:

$$F = \rho_p a A_b I_{sp} p^n \tag{3.12}$$

where ρ_p is the density of the solid propellant, and A_b is the area of the burning surface. From the result, thrust is increased with burning rate. As mentioned in the previous section, grain geometry affects the thrust profile. For progressive or regressive thrust profile, A_b is enlarged or shrank during combustion, so that thrust is increased or decreased, respectively.

Stable combustion requires $n < 1$, which will be proven here. When the nozzle is choked in the throat, the mass flow rate $\dot{m}_e$ is shown as:

$$\dot{m}_e = \frac{p A_t \Gamma}{\sqrt{RT}} \tag{3.13}$$

$$\Gamma = \sqrt{\gamma \left(\frac{2}{\gamma + 1} \right)^{\frac{\gamma - 1}{\gamma + 1}}} \tag{3.14}$$

where A_t is the throat area, γ is the specific heat ratio, R is the gas constant, and T is the gas temperature in the thrust chamber. The conservative law of mass is:

$$\frac{\partial}{\partial t} \int_{V_g} \rho dV + \int_{CS} \rho \left(\vec{u} \cdot \vec{n} \right) dA = 0 \tag{3.15}$$

where V_g is the gas region in the thrust chamber, which is a function of time because the solid propellant regresses during combustion. A is the surface on volume V_g, $\overrightarrow{n}$ is the unit normal vector of A, and ρ is the gas phase density. The term CS stands for the control surface, which covers V_g. The first term on the left-hand side is rewritten as:

$$\frac{\partial}{\partial t}\int_{V_g}\rho dV = \rho A_b r + V_g\frac{\partial\rho}{\partial t} \tag{3.16}$$

The second term on the left-hand side expresses the mass flow rates passing out through the surface A and then written as:

$$\int_{CS}\rho\left(\overrightarrow{u}\cdot\overrightarrow{n}\right)dA = \dot{m}_e - \rho_p A_b r \tag{3.17}$$

So that, from $\rho_p \gg \rho$:

$$V_g\frac{\partial\rho}{\partial t} + \frac{A_t\Gamma}{\sqrt{RT}}p - \rho_p A_b r = 0 \tag{3.18}$$

Hence, the equilibrium pressure in the steady-state p_{eq} is expressed using equilibrium temperature T_{eq} as:

$$p_{eq} = \left(\frac{\Gamma A_t}{\rho_p a A_b\sqrt{RT_{eq}}}\right)^{\frac{1}{n-1}} \tag{3.19}$$

Then, let us consider the combustion stability, i.e., the stability of the solution p_{eq}. In other words, p_{eq} shows neither oscillation nor diverge. For simplicity, the polytropic process (polytropic index: δ) is assumed because the relation of variables is expressed using simple equations: $p/\rho^\delta = p_{eq}/\rho_{eq}^\delta$ and $p/T^{\frac{\delta}{\delta-1}} = p_{eq}/T_{eq}^{\frac{\delta}{\delta-1}}$. Eq. (3.18) is a nonlinear differential equation, and there is no analytical solution. Then, Eq. (3.18) is linearized using:

$$X = X_{eq} + \widetilde{X} \tag{3.20}$$

where $\widetilde{X}$ is the small variation, and X can be p, ρ, and T. Linearizing the polytropic process relations using $(1 + \widetilde{X})^n \approx 1 + n\widetilde{X}$ (for $\widetilde{X} \ll 1$) and substitution into Eq. (3.18) lead:

$$\frac{d\widetilde{p}}{dt} = \frac{A_t\delta\Gamma\sqrt{RT_{eq}}}{V_g}\left(n - \frac{\delta+1}{2\delta}\right)\widetilde{p} \tag{3.21}$$

In order that the solution is asymptotically stable, i.e., $t \to \infty$ leads $\widetilde{p} \to 0$, the following condition should be met:

$$n < \frac{\delta+1}{2\delta} \tag{3.22}$$

If the process is the isothermal process, we can use $\delta = 1$ and gain:

$$n < 1 \tag{3.23}$$

From the discussion, for the thruster using steady combustion, the solid propellant should meet $n < 1$. Note that the stable is the local stable, and hence, when the pressure variation is large, thrust chamber pressure can oscillate in the way of the limit cycle. Moreover, the values a and n are dependent on the thrust chamber pressure range for a certain solid propellant. From the discussion, $n < 1$ is the minimum required condition for stable thrust production.

3.3.2 Combustion model

This chapter discusses the balance of energy. Here, D denotes the substantial differentiation in the Lagrangian description. The equation of specific enthalpy h conservation is expressed as:

$$\rho \frac{Dh}{Dt} = \frac{Dp}{Dt} - \nabla \cdot \vec{q} + \Phi + \dot{Q}_R + \rho \sum Y_k \vec{f}_k \cdot \vec{u}_k \tag{3.24}$$

where $\vec{q}$ is the heat flux through the conduction, Φ is the dissipation function, which is the kinetic energy loss due to the viscosity, and $\dot{Q}_R$ is the radiative heat addition to the fluid per unit volume [W/m^3]. Y_k, $\vec{u}_k$, and $\vec{f}_k$ is the mass fraction, diffusion velocity, and external volumetric force of the k-th reactant, respectively. For simplicity, let us assume that the combustion gas has neither viscosity, heat transfer, nor volumetric forces $\vec{f}_k$ and that the reaction proceeds at the constant pressure. Then, the equation is reduced to:

$$\rho \frac{Dh}{Dt} = - \nabla \cdot \vec{q} \tag{3.25}$$

The enthalpy h is expressed using the enthalpy of formation for i-th reactant $\Delta h^\circ_{f,i}$ and specific heat at constant pressure $C_{p,i}$, which is the function of temperature for the ideal gas:

$$h = \sum Y_i h_i \tag{3.26}$$

$$h_i = \Delta h^\circ_{f,i} + \int C_{p,i} dT \tag{3.27}$$

Using thermal conductivity λ, Fourier's law is:

$$\vec{q} = - \lambda \nabla T \tag{3.28}$$

Assuming that the thermal conductivity is isotropic and $C_{p,i}$ are the uniform, we gain:

$$\rho C_p \frac{DT}{Dt} = \lambda \frac{\partial^2 T}{\partial x^2} - \sum \Delta h^\circ_{f,i} \omega_i \tag{3.29}$$

where ω_i is reaction mass rate per unit volume [kg/m^{3}s]. For simplicity, the heat of reaction Q [J/kg] and reaction mass rate per unit volume ω [kg/m^{3}s] are applied, so that:

$$\rho C_p \frac{DT}{Dt} = \lambda \frac{\partial^2 T}{\partial x^2} + \omega Q \tag{3.30}$$

The equation is expressed in the Eulerian description:

$$\rho C_p \frac{\partial T}{\partial t} + \rho C_p u \frac{\partial T}{\partial x} = \lambda \frac{\partial^2 T}{\partial x^2} + \omega Q \tag{3.31}$$

where flow velocity u is equal to the burning rate r and flow velocity for the solid and gas phases, respectively. In other words, for simplicity, we can interpret that the solid propellant moves toward $+x$ direction at the velocity of r such that surface position is fixed at $x = 0$, whereas we have images that the burning surface regresses.

The quantities Q and ω are the function of pressure and temperature and dependent on solid propellant. For HTPB/AP composite propellants, HTPB and AP are decomposed and gasified on the burning surface, and their reaction is endothermic. Hence, $Q < 0$ at $x = 0$ and $Q = 0$ for $x < 0$. Decomposition on the burning surface is sustained by the heat flux from the flame. Then, the decomposed gases diffuse and are mixed in the gas phase, so that heat is released by combustion. Therefore $Q \geq 0$ for $x > 0$. Note that the combustion process is dependent on the thrust chamber pressure and AP diameter; premix combustion is dominant at lower thrust chamber pressure and diffusive combustion for higher thrust chamber pressures. In contrast, for DB propellants, the chemical reaction on the burning surface is exothermic, so that $Q > 0$ at $x = 0$.

Next, let us consider the temperature profile of the solid phase. Interestingly, although the burning surface temperature exceeds 700 K, the temperature exponentially decreases with the distance from the burning surface (decreasing x) so that temperature in most of the solid phase is as low as that before ignition. The heated region is restricted to the neighboring region, whose thickness is less than 1 mm. This can be explained using Eq. (3.31).

For the steady-state, Eq. (3.31) in the solid phase is reduced to:

$$\rho C_p u \frac{dT}{dx} = \lambda \frac{d^2 T}{dx^2} \tag{3.32}$$

because no chemical reaction induced in the solid phase except the burning surface, so that $\omega = 0$ for $x < 0$. Integrating Eq. (3.32) with x using the boundary conditions ($T = T\infty$ ($x = -\infty$) and $T = T_s$ ($x = 0$)) yields:

$$T = T_\infty + (T_S - T_\infty)\exp\left(\frac{\rho_p C_p r}{\lambda}x\right) \quad (x \leq 0) \tag{3.33}$$

From Eq. (3.33), temperature exponentially decreases with distance from the burning surface. Hence, in the solid-propellant propulsion, the propellants play a role of heat insulators that prevent heat conduction from flames to protect the thruster casing from high-temperature gas. Heat penetration thickness δ is defined as:

$$\delta = \frac{\lambda}{\rho_p C_p r} \tag{3.34}$$

The quantity δ expresses the penetration depth of heat produced by combustion. The heat penetration is thinned with increasing burning rate r.

3.4 Design of solid-propellant thruster

In designing solid-propellant propulsion, we need to determine throat area A_t, expansion ratio ε, grain geometry, and thrust chamber volume. The chapter addresses the design procedure. First of all, we need to evaluate the theoretical performance of solid-propellant thrusters. Then, specific impulse, combustion products, thrust chamber temperature T_c, Mach number, and sonic velocity are calculated using chemical equilibrium calculation for a certain thrust chamber pressure p_c, expansion ratio, propellant ingredients, and mixture ratios. Then, the target mass flow rate of propellant $\dot{m}$ can be determined from I_{sp}, target thrust F using the definition of I_{sp}:

$$\dot{m} = \frac{F}{g I_{sp}} \tag{3.35}$$

Since the nozzle is usually choked during the thruster firing, the mass flow rate is related to the density ρ_t and sonic velocity a_t, which was evaluated by the chemical equilibrium calculation. The throat diameter is determined from:

$$A_t = \frac{\dot{m}}{\rho_t a_t} \tag{3.36}$$

Then, we can determine burning surface area A_b using burning rate r and propellant mass density ρ_p:

$$A_b = \frac{\dot{m}}{\rho_p r} \tag{3.37}$$

If the quantities a and n in Vieille's law are known:

$$A_b = \frac{\dot{m}}{\rho_p a p_c^n} \tag{3.38}$$

Theoretical specific impluses are evaluable using chemical equilibrium calculation. NASA released Chemical Equilibrium Application (CEA), which bundles command-prompt-based software of chemical equilibrium calculation and JAVA-based graphical user interface [23]. However, the availability is degraded now for security reasons. However, a web-based simulator, the CEA run, is released on the net. RocketCEA is another option, which uses Python. Combining plot libraries on Python allows us to readily visualize the performance dependence, such as the dependence of specific impulse on mixture ratio [24].

However, the determination of thrust chamber volumes is not as easy as that for A_t. In general, the volume is characterized using characteristic chamber length L^*, which is defined as:

$$L^* = \frac{V_g}{A_t} \tag{3.39}$$

In general, L^* should be sufficiently large to complete combustion processes inside thrust chambers, but enlarging L^* lowers thrust and specific impulse because increasing the heat loss through thrust chamber walls. Moreover, time variation in thrust chamber pressure becomes moderate by enlarging L^*. Hence, L^* needs to be empirically determined.

3.5 Progress in solid-propellant microthruster

This chapter introduces the research examples of developed solid-propellant microthrusters. Today, many thrusters are developed by universities and research institutes, as listed in Table 3.2. MEMS are actively introduced to the fabrication, and some have arrays of miniaturized thrusters. Moreover, their ignition devices and propellants are different from conventional solid-propellant engines. Hence, solid-propellant microthrusters are brand-new rather than miniaturized solid rocket engines.

3.5.1 Non-MEMS microthruster

Laser ignition microthrusters using B/KNO_3 propellant have been developed to provide pulsed thrust. B/KNO_3, which yields low specific impulse and hence has been usually used as pyrolant, was applied because the propellant is readily ignited in a vacuum. The pyrolant effectively absorbs near-infrared beams, and hence, laser diodes, which are able to emit high-power near-infrared beams with high efficiency, were used as a heat source. Moreover, the use of focusing optical lenses allows laser heating with higher heat flux. Koizumi et al. designed a dual-mode microthruster illustrated in Fig. 3.14 and showed that the thruster provided an impulsive thrust of 11.2 ± 2 mNs with an I_{sp} of 36 ± 7 s [25]. The thruster yielded a higher specific impulse with a lower impulsive thrust using laser ablation of polyacetal added with carbon black, which was added to absorb the laser beam.

Nakano et al. developed a laser ignition B/KNO_3 microthruster (Fig. 3.15) to alter the attitude of Kiseki (KKS-1) microsatellite. The thruster yielded an impulsive thrust of $40-70$ mNs with a specific impulse of 100 s. KKS-1 spacecraft was launched on January 23, 2009, by H-IIA rocket No. 15 and successfully inserted into the Earth orbit [45]. Asakawa et al. developed a higher ΔV microthruster where

Table 3.2 List of solid-propellant microthrusters.

Ref	Fabrication	Array	Propellant	Ignitor	Thrust, mN	Impulse, mNs	I_{sp}; s	Note
[25]	Machining	8 × 3 (cylindrical)	B/KNO$_3$	Laser diode	-	11 ± 2	36 ± 7	Dual mode
[26]	Machining	5 × 1 (round)	B/KNO$_3$	Laser diode	-	40–70	100	
[27]	Machining	Single	B/KNO$_3$	Laser diode	22,900–45,500	17,500–24,600	160–193	
[28]	Machining	Single	HTPB/AP/C	Laser diode	60	Adjustable	144	Repetitive firing
[29]	Machining	6 × 6	GAP/AP	Resistive igniter wire (Ni/Cr)	0.021–0.29	0.04–0.379	0.42–4.1	
[30]	Machining	Single	Nano Al/CuO	Heater	30,000–70,000 (peak)	2–5.2	20–25	
[31]	Machining	Single	Al, Bi$_2$O$_3$/CuO/NC	Heater	-	4.6–46	20–60	
[32]	MEMS, vertical	15 × 15	AP-based composite	Heater	-	-	-	
[33]	MEMS, vertical	4 × 4	GAP/AP	Heater with ZPP	0.36, 2.3 (peak)	0.135, 7.3	-	
[34]	MEMS, vertical	10 × 10	B/KNO$_3$	Heater with pyrolant	-	0.02–0.3	-	
[35]	MEMS, vertical	22	B/KNO$_3$	Heater w/B/Ti pyrolant	-	0.5	-	
[36]	MEMS, planar	7 × 1	DB + GP	Polysilicon heater	0.1–1 (average)	0.17–1.13	2.48 −16.5	
[33]	MEMS, vertical	4 × 4	GAP/AP/Zr	Heater with ZPP	0.36, 2.3 (peak)	0.135, 7.3	-	
[37]	MEMS, vertical	10 × 10	GAP/AP/Zr	Heater with ZPP	0.3–2.3	-	-	

Continued

Table 3.2 List of solid-propellant microthrusters.—*cont'd*

Ref	Fabrication	Array	Propellant	Ignitor	Thrust, mN	Impulse, mNs	I_{sp}; s	Note
[38]	MEMS, vertical	10 × 10 100 × 100	HTPB/AP	Heater	-	0.2125	23.4	
[39]	MEMS, planar	5 × 1	GP/AP/Al/Fe$_2$O$_3$	Au/Ti heater	-	0.0352–0.222	4.48 −28.29	
[40]	MEMS, planar	Single	LTNR/Nitro cotton	Heater	25 (peak)	6.3	-	
[41]	MEMS, vertical	3 × 5	LTNR	Polysilicon heater	-	0.09–0.12	-	
[42]	MEMS, vertical	10 × 10	Al/CuO/NC	Polysilicon heater	479–645	0.1559–0.346	10.2 −27.2	
[43]	MEMS, vertical	2	ZPP	Polysilicon heater	-	-	-	
[44]	MEMS, planar	5 × 7	HTPB/AP	Polysilicon resistive heater	10–15 peak	-	14	

GP, *Gunpowder.*

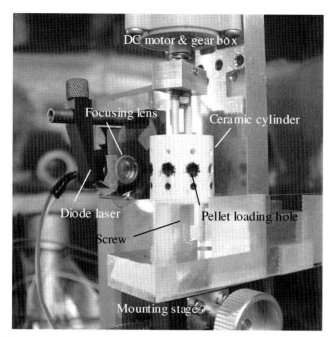

FIGURE 3.14

Dual-mode B/KNO$_3$ thruster using laser ignition.

B/KNO$_3$ pellets were stacked, as illustrated in Fig. 3.16. Thrust measurement gave a thrust impulse, average thrust, and specific impulse of 175−246 Ns, 22.9−45.5 N, and 160−193 s, respectively [27].

In other research, laser diodes are used as a heat source for switching on/off combustion to provide the capability of interrupting and restarting thrust production.

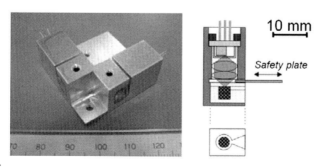

FIGURE 3.15

B/KNO$_3$ thruster using laser ignition for KKS-1 microsatellite [45].

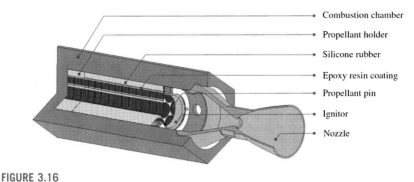

FIGURE 3.16

B/KNO$_3$ thruster using laser ignition for high ΔV microsatellite [27].

Conventional solid-propellant propulsion generally has difficulties in interrupting and restarting combustion since the propellant keeps burning once ignited. Hence, Isakari et al. developed the solid propellant that kept burning during laser heating on the burning surface and interrupted combustion by swithing off laser heating. The combustion-controllable solid propellant is applied to microthrusters. Then, the thrust production was controllable by switching on and off laser heating [28]. Fig. 3.17 illustrates a 100-mN prototype with HTPB/AP/C = 30/70/0.5wt%. The theoretical specific impulse was 204 s (frozen flow, expansion ratio of 50, thrust chamber pressure of 0.1 MPa). At 15-W laser heating, the thruster started firing with ignition delays ranging from 3 to 5 s and sustained stable combustion during laser heating. Switching off laser emission interrupted combustion. Thrust

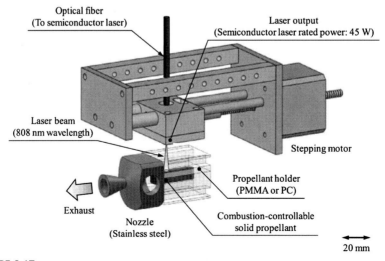

FIGURE 3.17

Combustion controllable HTPB/AP microthruster using laser-assisted combustion [28].

measurement showed that the maximum thrust and specific impulse was 50 mN and 144 s with an ignition delay of 3.3 s.

The thermite reaction was applied to some microthrusters. Staley et al. added NC to the nanothermites [31]. The prototype was examined using Al/CuO/NC and Bi_2O_3/Al/NC with NC mixture ratios of 2.5 and 5.0wt%. Al and Bi_2O_3 are nanospheres 90−210 and 80 nm in diameter, respectively, and CuO is a nanorod 8−12 nm in diameter and 100 nm in length. Thrust measurement showed that the thruster yielded specific impulses ranging from 50 to 55 s, and TMD had an insignificant influence on specific impulse. For Bi_2O_3/Al/NC, an increase in NC mixture ratio enlarged specific impulse, which reached the maximum value of 59.4 ± 2.9 s at 5.0wt%. In contrast, the test with Al/CuO/NC presented a maximum specific impulse (24.2 ± 3.8 s) with a volumetric specific impulse (0.6 ± 0.1 mN/mm^3) at 2.5% NC mixture ratio.

Sathiyanathan et al. prototyped a GAP/AP microthruster [29]. This group machined a 4×1 array type prototype with a 40-gauge nichrome wire heater for ignition. In the thruster, the thrust chamber and de Laval nozzle are vertically piled in the same way as the MEMS-based thruster, which will be introduced in the following section. The thrust measurement showed that the prototype produced 0.04−0.379 mNs thrust with 0.42−4.1 s specific impulses.

3.5.2 MEMS-based microthruster

Besides these machined microthrusters, some are fabricated using MEMS, which is the precise machining technology for semiconductor fabrication. There are two types of structure: planar and vertical types. In the former type, the thrust chamber and nozzle are formed horizontally, and in the latter, the thrust chamber and nozzles are piled up vertically. Moreover, the miniaturized rockets are arrayed in a MEMS-based thruster to produce multishot thrust impulses, and those to be fired were selected using the row-column addressing control method, which is used in the random access memory (RAM) for computers. In such thrusters, the miniaturized rockets are located like a matrix, and the ignitors (usually resistive heaters) are connected to the row and column lines. The rocket to be ignited is selected by the memory addressing (selecting the row and column lines one by one and applying a voltage to the two lines). Then, let us focus on the MEMS-based thrusters developed by institutes.

In 2000, Lewis et al. designed a 3×5-array thruster using LTNR (explosive) as the propellant [41]. This is the vertical type having a RAM-like structure with resistive heaters connected to row/column lines. The heater was a 210-Ω polysilicon resistor formed on a SiO_3 insulation layer 3-μm in thickness and consumed 100-W power and 100 mJ energy. The rupturable diaphragms prevented the propellant leak in standby and were broken at ignition so that the combustion gas was exhausted through the nozzles. Thrust measurement showed that the thruster provided thrusts ranging from 0.09 to 0.12 mN.

Moreover, Shen et al. fabricated a microthruster with LTNR/Nitro cotton = 50/50 wt% (Fig. 3.18) [40]. These were a planer type 8.8 mm long, 6.0 mm wide, and

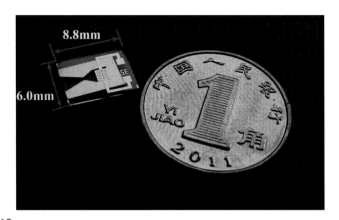

FIGURE 3.18

LTNR microthruster [40].

1.5 mm high and were made of Pyrex glass and silicon. The ignitor heater was chromium 370 nm thick on the silicon and had an average resistance of 140 Ω. The thrust impulse was 6.3 mNs with a peak thrust of 25 mN.

The GAP, which is an energetic material, was also applied to microthrusters. Rossi et al. designed a vertical type thruster (4 × 4 arrayed) with GAP/AP/Zr propellant (Fig. 3.19) [33]. The thrust chamber ($1.5 \times 1.5 \times 0.34$ mm^3) was filled with the propellant and pyrolant. The thrust chamber and nozzle were sealed with dielectric membranes (SiN_x(0.6 µm), SiO_2(1.4 µm)) so that the pyrolants and propellants were not exposed to the vacuum. The ignitor was formed on the dielectric membranes and used a polysilicon resistive heater ($450 \times 450 \times 0.5$ µm^3, $720 \times 720 \times 0.5$ µm^3) with pyrolants (ZPP or GAP/AP/Zr) for ensuring ignition.

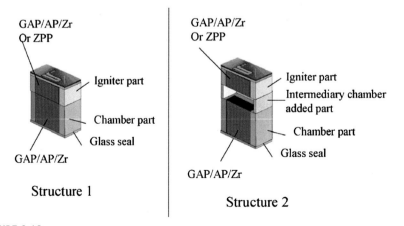

FIGURE 3.19

GAP/AP/Zr propellant microthruster [33].

There were two prototypes: Structure 1 and 2; in the former, the propellant and pyrolant were contacted without any gap, while in the latter, the pyrolant and propellant were separated by the intermediary chambers. Firing tests showed that Structure 1 was not started in the case with the GAP/AP/Zr propellant. Structure 2 with ZPP pyrolant provided a 100% ignition rate at only 17-mJ heater energy, whereas that with GAP/AP/Zr pyrolant yielded an ignition rate of 50% at 73-mJ heater energy. The thrust measurements showed that the thrust impulse was 7.3 mNs for the ZPP pyrolant. This group also prototyped 10 × 10 arrayed thrusters where Zenner polysilicon diodes were applied for the addressing of heaters and showed the thruster with Foturan thrust chamber yielded 0.3−2.3 mNs [37].

Some microthrusters utilize conventional pyrolants as propellant. Briand et al. formed vertical-type thruster, the structure of which was almost the same as those shown in Ref. [33]. As presented in Fig. 3.20, to ensure stable ignition and combustion, ZPP, which is an energetic material that ignites at 200−300°C, was selected as the propellant [43]. The ignitor used a polysilicon heater where 2.8 kΩ beams 20 μm wide and 180 μm long were placed in parallel. Heating the propellant at 80 mW/beam induced combustion with an ignition delay of 120 ms (Fig. 3.21).

The B/KNO_3 pyrolant was applied to the MEMS-based microthruster either. Tanaka et al. prototyped a 10×10-arrayed vertical type [34]. A Pt/Ti resistive heater was used for ignition and was placed on the diaphragm that firing ruptured so that the combustion gas was expelled from the nozzle. To increase the ignition rate, they tested the Pt/Ti heater with lead rhodanide/potassium chlorate/NC. Without the

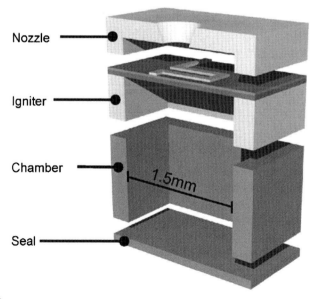

FIGURE 3.20

ZPP propellant microthruster [43].

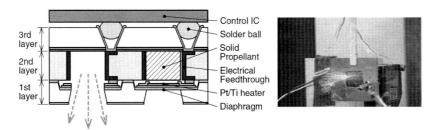

FIGURE 3.21

B/KNO$_3$ microthruster [34].

pyrolant, the ignition required 4−6 W heater power, but the use of the pyrolant reduced the heater power to 3−4 W. Moreover, the use of the pyrolant reduced ignition delay from 200 to 1000 ms (w/o pyrolant) to 12−30 ms. Thrust ranged from 0.02 to 0.3 mNs.

Gunpowder was also utilized as the propellant. Zhang et al. fabricated a gunpowder-based 5 × 1 arrayed planar-type microthruster with a mixture ratio of gun powder (75% KNO$_3$, 15% charcoal, 10% sulfur)/AP/Al/Fe$_2$O$_3$ = 100/6/3/1wt% [39]. As shown in Fig. 3.22, the prototype had five miniaturized thrusters and an ignitor wherein a 77-nm thick gold conductor was used as a lead line, and a 206-nm thick Ti resistor is used as the heater and adhesion layer. At a heater power of no less than 0.16 W, the thruster was fired but yielded an ignition delay of 12.9 s and corresponding ignition energy of no less than 2.07 J. The thrust impulse ranged from 0.0352 to 0.0222 mNs with vacuum specific impulses ranging from 4.48 to 28.29 s.

The HTPB/AP composite solid propellant is applied to the MEMS-based microthrusters. Teasdale et al. fabricated a 5×7-arrayed planar-type thruster (Fig. 3.23) [44]. The ignitor used a p- or n-type polysilicon resistive heater. The p-type

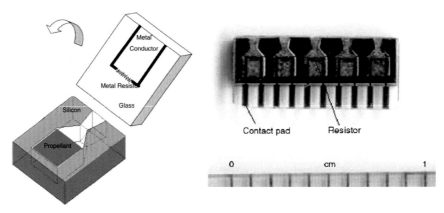

FIGURE 3.22

Gun powder microthruster with Au/Ti ignitor [39].

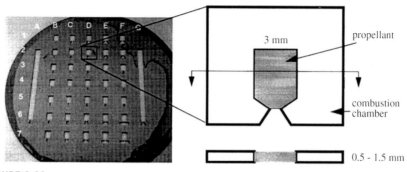

FIGURE 3.23

HTPB/AP microthruster using polysilicon heater [44].

polysilicon heater had a thickness of 0.6 μm, a resistance of 1.1 kΩ, and electric power of 0.3 W, and the n-type polysilicon heater had 0.6 μm, 1.1 kΩ, and 0.2 W. The firing test showed that the thruster was ignited by 4−300 mJ heater energy with ignition delays ranging from 0.02 to 1.5 s. The specific impulse was 14 s with a peak thrust of 10−15 mN.

Liu et al. prototyped the 10×10 and 100×100 vertical-type thrusters illustrated in Fig. 3.24 [38]. The propellant is particulate HTPB/AP 100−200 μm in diameter. For ignition, the prototype used resistive heaters, and BPs and self-made ignition powder (the mixture of BP, red phosphorus, and KNO_3) were also used as the pyrolant. The pyrolants 50−100 μm and 100−200 μm in diameter were examined for increasing the ignition reliability. The ignitor tests showed that whereas the thruster with the conventional BP yielded ignition rates ranging from 10.0% to 23.3%, the thruster with the self-made ignition powder provided 100%

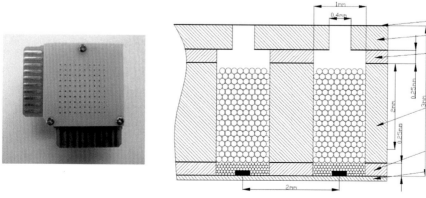

FIGURE 3.24

HTPB/AP microthruster [38].

and 80% ignition rates for 50–100 μm and 100–200 μm diameters, respectively. The required power was 0.72 W. The tests with HTPB/AP propellant presented a 100% ignition rate at a 0.89-W heater power. The 10 × 10 arrayed thrusters yielded 0.0586–0.2125 mNs thrust impulse with an average specific impulse of 12.0 s.

The thermite reaction was applied to some MEMS-based microthrusters. Apperson et al. prototyped a thruster shown in Fig. 3.25 [30]. CuO (nanosphere 80 nm in average diameter) and Al (nanorod) were used as a propellant with a mixture ratio of Al/CuO = 17/40 wt%, at which the previous research had shown the maximum combustion ratio. The percentage of theoretical maximum density (%TMD) was used as a parameter. TMD is the ratio of mass density to the theoretical maximum density of Al/CuO (5.36×10^3 kg/m^3) and was varied from 28.0% to 64.9% in the tests. The nozzle-less prototype presented lower combustion periods with thrust peaks for low TMD% region, whereas it extended the combustion periods with relatively uniform thrusts beyond 44.4% TMD. Thrust impulse was increased from 2 to 5.5 mNs with increasing %TMD, and specific impulse was almost maintained in the range from 20 to 25 s. The prototype with a de Laval nozzle was tested and showed the thrust time history different from the nozzle-less type and yielded no increase in specific impulse.

Nanothermite propellants were also examined, and Ru et al. tested Al/CuO nanothermite with NC as the propellant of a vertical-type 10 × 10 arrayed prototype [42]. A microsemiconductor bridge that was fabricated using n-type heavily doped polysilicon was a resistive heater 12.7 Ω in resistance and 5 W in power consumption. The mixture ratio Al/CuO was fixed at 28.8/71.2wt%, while the NC mixture ratio was varied from 0 to 10wt%. Theoretical specific impulse increased with increasing NC mixture ratio; 73.6 s for 0% NC mixture ratio and 78.7 s for 10wt% NC. The electrospray and mechanical mixing were utilized to mix nano Al and nano CuO. All the prototypes were successfully fired except those with Al/CuO/NC(10wt%). For the use of mechanical mixing case, the thruster presented a specific impulse and thrust impulse of 10.2 ± 3.3 s, 0.1559 ± 0.0449 mNs at 0wt% NC, and the maximum values were 17.7 ± 2.5 s, 0.2552 ± 0.0318 mNs at 2.5wt% NC. On the other hand, for the use of the electrospray, specific impulse and thrust impulse were monotonically increased with increasing NC mixture ratio with the maximum values of 27.2 ± 4.4 s, 0.3469 ± 0.0562 mNs at 10wt% NC (Fig. 3.26).

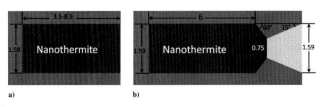

FIGURE 3.25

CuO/Al nanothermite microthruster [30].

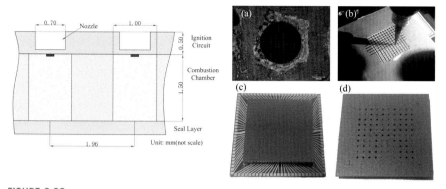

FIGURE 3.26

Nanothermite 10×10 arrayed MEMS thruster [42].

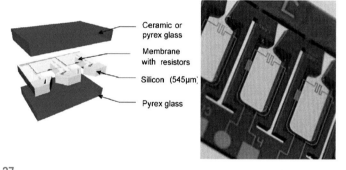

FIGURE 3.27

DB/GP microthruster [36].

Chaalane et al. tested the double-base propellant with BP [36]. Fig. 3.27 illustrates a planar-type 7×1 arrayed thruster. A resistive heater, which was formed on SiO_2/SiN_x membrane, was 540×540 μm^2 polysilicon and consumed 600-mW electric power. The mixture ratio of BP was varied from 0 to 30wt%. The firing tests showed that the prototype produced combustion for 10, 20, and 30wt% BP with an ignition delay of 150 ms while no ignition was induced for 0wt% BP. Average thrust and thrust impulse were increased with increasing BP mixture ratio and reached the maximum values of 1 mN and 1.13 ms. The propellant mass was approximately 7 mg so that the specific impulse was approximately 16 s.

From these research examples, there are many novel and ambitious efforts in developing solid-propellant microthrusters. The solid-propellant microthrusters are more diverse in the propellant and manufacturing methods than the conventional solid rockets used in the launching vehicle.

3.6 Conclusion and future prospects

The chapter addressed the fundamentals and principles of solid-propellant propulsion, including the propellants and the features. Besides machined microthrusters inheriting the conventional structure, integrated-circuit-like thrusters were developed with MEMS technology. Whereas conventional propellants such as HTPB/AP and GAP were tested, some groups examined pyrolants and explosives to ensure ignition under vacuum conditions with lower ignition delay. Today, small satellites are now going to have the thruster to expand the mission fields, and solid-propellant propulsion is one of the promising options because of the simplicity and higher impulse capability.

In the future, solid-propellant microthrusters would enhance diversity compared with the launching and sounding solid-propellant rocket engines because spacecraft are used in various missions, and even low thrust and short-duration firing make sense in some nanosatellites. The technologies that have been proposed but not or rarely applied to the practical solid-propellant rocket engines may be introduced to solid micropropulsion. A candidate may be the pintles, which have been tested in the large-scale solid rocket to realize the variable thrust [46]. A needle-like pintle, placed in the thrust chamber, is inserted in the nozzle so that the effective area of the nozzle throat is adjusted to vary thrust actively and flexibly. Thrust termination valves and variable-area nozzles could be used either. MEMS technology has the potential to sophisticate such old but attractive technologies and allows them to be implemented in micropropulsion.

Conventional-shape solid microthrusters could gain benefits by the application of MEMS-based parts. MEMS technology can create small parts such as ignitors and laser diodes, which are distributable to thruster and enable precise combustion control. The collaboration of conventionally-structured thruster and new MEMS would also provide new valuable features of onboard solid-propellant propulsion.

As mentioned in the introduction, the solid-propellant rocket engines were the first onboard device but were phased out; instead, liquid-propellant thrusters have dominated the spacecraft propulsion. The new solid-propellant microthruster can change the game owing to the compactness, simplicity, and relatively high performance. In the future, we may see the solid-propellant propulsion resurrect in spacecraft to vigorously propel nanosatellites as if a stone once rejected revived as the cornerstone.

References

[1] Technical Report NASAMEMO-5-25-59W/vol. 3 1958 NASA/USAF Space Probes (ABLE-l) Final Report, Vol. 3, June 1959. NASA.
[2] The Marin Company, April 1960. The Vanguard Satellite Launching Vehicle an Engineering Summary. Technical Report 11022 (Engineering Report).

[3] Technical Report NASAMEMO-5-25-59W/vol. 2. 1958 NASA/USAF Space Probes (ABLE-l) Final Report, Vol. 2, June 1959. NASA.

[4] Naminoskuke Kubota, 2007. Propellants and Explosives, second ed. WILEY-VCH Verlag GmbH & Co. KGaA, Weinheim.

[5] Schoeyer, H.F.R., Schnorhk, A.J., Korting, P.A.O.G., van Lit, P.J., Mul, J.M., Gadiot, G.M.H.J.L., Meulenbrugge, J.J., 1995. High-performance propellants based on hydrazinium nitroformate. J. Propul. Power 11 (4), 856−869.

[6] Eamon Colclough, M., Desai, H., Millar, R.W., Paul, N.C., Stewart, M.J., Golding, P., 1994. Energetic polymers as binders in composite propellants and explosives. Polym. Adv. Technol. 5 (9), 554−560.

[7] Frankel, M.B., Grant, L.R., Flanagan, J.E., 1992. Historical development of glycidyl azide polymer. J. Propul. Power 8 (3), 560−563.

[8] Wingborg, N., 2019. Heat of formation of ADN-based liquid monopropellants. Propellants, Explos. Pyrotech. 44 (9), 1090−1095.

[9] Khare, P., Yang, V., Meng, H., Risha, G.A., Yetter, R.A., 2015. Thermal and electrolytic decomposition and ignition of HAN-water solutions. Combust. Sci. Technol. 187 (7), 1065−1078.

[10] National Center for Biotechnology Information, 2020. Pubchem Compound Summary for Cid 4510, Nitroglycerin. Retrieved November 5, 2020. https://pubchem.ncbi.nlm.nih.gov/compound/Nitrocellulose.

[11] Volk, F., Bathelt, H., 2002. Performance parameters of explosives: equilibrium and nonequilibrium reactions. Propellants, Explos. Pyrotech. 27 (3), 136−141.

[12] Agrawal, J.P., 2010. High Energy Materials. 2010 WILEY-VCH Verlag GmbH & Co. KGaA, Weinheim.

[13] Krien, G., Licht, H.H., Zierath, J., 1973. Thermochemical investigation of nitramines. Acta Thermochem. (6).

[14] Kumar, P., 2018. An overview on properties, thermal decomposition, and combustion behavior of adn and adn based solid propellants. Defence Technol. 14 (6), 661−673.

[15] Anflo, K., Mllerberg, R., 2009. Flight demonstration of new thruster and green propellant technology on the PRISMA satellite. Acta Astronaut. 65 (9), 1238−1249.

[16] Hori, K., Katsumi, T., Sawai, S., Azuma, N., Hatai, K., Nakatsuka, J., 2019. Han-based green propellant, SHP163 its R&D and test in space. Propellants, Explos. Pyrotech. 44 (9), 1080−1083.

[17] Spores R.A., 2015. GPIM AF-M315E Propulsion System. In: 51st AIAA/SAE/ASEE Joint Propulsion Conference.

[18] National center for biotechnology information, 2020. Pubchem Compound Summary for Cid 4510, Nitroglycerin. Retrieved November 5, 2020. https://pubchem.ncbi.nlm.nih.gov/compound/Nitroglycerin.

[19] McKeen, L.W., 2007. Permeability Properties of Plastics and Elastomers, third ed. Elsevier, Amsterdam.

[20] Jessup, R.S., Prosen, E., 1950. Heats of combustion and formation of cellulose and nitrocellulose (cellulose nitrate). J. Res. Natl. Bur. Stand. 44, 387.

[21] Freedman, E., July 1982. Blake - a Thermodynamics Code Based on Tiger: Users' Guide and Manual. Technical Report ARBRL-TR-12411.

[22] Yano, Y., 1989. Condensed phase reaction of boron with potassium nitrate. Propellants, Explos. Pyrotech. 14 (5), 187−189.

[23] https://cearun.grc.nasa.gov/.

[24] https://rocketcea.readthedocs.io.

[25] Koizumi, H., Inoue, T., Arakawa, Y., Nakano, M., 2005. Dual propulsive mode micro-thruster using a diode laser. J. Propul. Power 21 (6), 1133−1136.

[26] Nakano, M., Koizumi, H., Watanabe, M., Arakawa, Y., 2009. A laser ignition micro-thruster for microspacecraft propulsion. Trans. Jpn. Soc. Aeronaut. Space Sci, Space Technol. Jpn. 7 (ists26), Pb7−Pb9.

[27] Asakawa, J., Koizumi, H., Kojima, S., Nakano, M., Komurasaki, K., 2019. Laser-ignited micromotor using multiple stacked solid propellant pellets. J. Propul. Power 35 (1), 41−53.

[28] Isakari, S., Asakura, T., Haraguchi, D., Yano, Y., Kakami, A., 2017. Performance evaluation and thermography of solid-propellant microthrusters with laser-based throttling. Aero. Sci. Technol. 71, 99−108.

[29] Sathiyanathan, K., Lee, R., Chesser, H., Dubois, C., Stowe, R., Farinaccio, R., Ringuette, S., 2011. Solid propellant microthruster design for nanosatellite applications. J. Propul. Power 27 (6), 1288−1294.

[30] Apperson, S.J., Bezmelnitsyn, A.V., Thiruvengadathan, R., Gangopadhyay, K., Gangopadhyay, S., Balas, W.A., Anderson, P.E., Nicolich, S.M., 2009. Characterization of nanothermite material for solid-fuel microthruster applications. J. Propul. Power 25 (5), 1086−1091.

[31] Staley, C.S., Raymond, K.E., Thiruvengadathan, R., Apperson, S.J., Gangopadhyay, K., Swaszek, S.M., Taylor, R.J., Gangopadhyay, S., 2013. Fast-impulse nanothermite solid-propellant miniaturized thrusters. J. Propul. Power 29 (6), 1400−1409.

[32] Rossi, C., Orieux, S., Larangot, B., Do Conto, T., Estve, D., 2002. Design, fabrication and modeling of solid propellant microrocket-application to micropropulsion. Sensor Actuator Phys. 99 (1), 125−133. Special issue from the papers presented in Symposium J in E-MRS 2001 conference.

[33] Rossi, C., Larangot, B., Lagrange, D., Chaalane, A., 2005. Final characterizations of MEMS-based pyrotechnical microthrusters. Sensor Actuator Phys. 121 (2), 508−514.

[34] Tanaka, S., Hosokawa, R., Tokudome, S., Hori, K., Saito, H., Esashi, M., 2003. MEMS-based solid propellant rocket array thruster. Trans. Jpn. Soc. Aeronaut. Space Sci. 46 (151), 47−51.

[35] Tanaka, S., Kondo, K., Habu, H., Itoh, A., Watanabe, M., Hori, K., Esashi, M., 2008. Test of B/Ti multilayer reactive igniters for a micro solid rocket array thruster. Sensor Actuator Phys. 144 (2), 361−366.

[36] Chaalane, A., Rossi, C., Estve, D., 2007. The formulation and testing of new solid propellant mixture (DB+x%BP) for a new MEMS-based microthruster. Sensor Actuator Phys. 138 (1), 161−166.

[37] Rossi, C., Briand, D., Dumonteuil, M., Camps, T., Pham, P.Q., de Rooij, N.F., 2006. Matrix of 10×10 addressed solid propellant microthrusters: review of the technologies. Sensor Actuator Phys. 126 (1), 241−252.

[38] Liu, X., Li, T., Li, Z., Ma, H., Fang, S., 2015. Design, fabrication and test of a solid propellant microthruster array by conventional precision machining. Sensor Actuator Phys. 236, 214−227.

[39] Zhang, K.L., Chou, S.K., Ang, S.S., Tang, X.S., 2005. A MEMS-based solid propellant microthruster with Au/Ti igniter. Sensor Actuator Phys. 122 (1), 113−123. SSSAMW 04.

[40] Shen, Q., Yuan, W., Li, X., Xie, J., Chang, H., 2015. An orthogonal analysis method for decoupling the nozzle geometrical parameters of microthrusters. Microsyst. Technol. 21 (6), 1157−1166.

[41] Lewis, D.H., Janson, S.W., Cohen, R.B., Antonsson, E.K., 2000. Digital micropropulsion. Sensor Actuator Phys. 80 (2), 143—154.

[42] Ru, C., Wang, F., Xu, J., Dai, J., Shen, Y., Ye, Y., Zhu, P., Shen, R., 2017. Superior performance of a mems-based solid propellant microthruster (spm) array with nanothermites. Microsyst. Technol. 23 (8), 3161—3174.

[43] Briand, D., Pham, P.Q., Nicolaas, F., de Rooij, 2007. Reliability of freestanding polysilicon microheaters to be used as igniters in solid propellant microthrusters. Sensor Actuator Phys. 135 (2), 329—336.

[44] Teasdale, D., Milanovic, V., Chang, P., Pister, K.S.J., 2001. Microrockets for smart dust. Smart Mater. Struct. 10 (6), 1145—1155.

[45] Nakano, M., Koizumi, H., Watanabe, M., Arakawa, Y., 2010. Laser ignition microthruster experiments on KKS-1. Trans. Jpn. Soc. Aeronaut. Space Sci., Aerospace Technol. Jpn. 8 (ists27), Pb7—Pb11.

[46] Levinsky, C.T., Kobalter, G.F., 1968. Feasibility Demonstration of a Single-Chamber Controllable Solid Rocket Motor. Technical Report AFRPL-TR-67-300.

Liquid propellant microthrusters

Kean How Cheah, PhD [1], **Wai Siong Chai, PhD** [2], **Toshiyuki Katsumi, PhD** [3]

[1]*Assistant Professor, School of Aerospace, Faculty of Science and Engineering, University of Nottingham Ningbo China, Ningbo, Zhejiang, China;* [2]*Post-doctoral Researcher, School of Mechanical Engineering and Automation, Harbin Institute of Technology, Shenzhen, Guangdong, China;* [3]*Associate Professor, Department of Mechanical Engineering, Nagaoka University of Technology, Nagaoka, Niigata, Japan*

4.1 Historical background and principles of operation

In the broadest classification of rocket propulsion, a chemical propulsion system is the one that generates thrust using chemical energy stored in the propellant, which could be in the physical state of solid, liquid, or a combination of both.

The historical background for the development of liquid propellant propulsion can be traced back to as early as the 19th century. It is well documented that the idea of liquid propellant propulsion had been proposed and examined independently in different parts of the world, almost concurrently. In Russia, Konstantin Tsiolkovsky (1857−1935) who is a mathematics teacher wrote about space travel, artificial satellites, and rocket propulsion. In 1903, he published an important work, named *Exploration of Outer Space by Means of Rocket Devices,* in which the famous Tsiolkovsky's equation was derived. Liquid oxygen and hydrogen have been identified as the suitable propellants in a multistage rocket to achieve a possible space flight. In Germany, Hermann Oberth (1894−1989) has published a book, named *The Rocket into Planetary Space*, based on his rejected doctoral dissertation, which details the similar idea of using liquid propellant rockets for space travel. The book has gained much attention among the amateur rocketry societies in Germany. He later became a member of the famous *Verein für Raumschiffahrt* (VfR) and inspired many young and enthusiastic members. One of them is Wernher von Braun (1912−77) who became the giant and leading figure in the development of German and American rocket and space technology.

Previous works by Tsiolkovsky and Oberth were mostly focused on mathematical theories of rocketry, until Robert H. Goddard (1882−1945), a US professor at Clark University, Massachusetts, who has studied not only the mathematical theories but also carried out extensive works on the designs and experiments of rocketry. On March 26, 1926, he became the first to build and launch a liquid-fueled rocket in Auburn, Massachusetts. Since then, he and his team has improved and launched multiple rockets between 1926 and 1941, during which the highest altitude of 2.6 km was recorded. Goddard has been widely credited with his notable contribution to the engineering aspects of rocketry, such as the use of gyroscope for steerable flight, delivery of propellant with turbopump, and cooling of the nozzle with liquid oxygen. Eventually, there are 214 patents granted to his rocketry works.

Today, almost all the major launch vehicles around the world are using liquid-fueled rocket engines as their first stage propulsion, supported by strapped on solid rocket boosters for additional thrust. These rockets are using liquid oxygen or dinitrogen tetroxide (N_2O_4) as oxidizer and liquid hydrogen, unsymmetrical dimethylhydrazine (UDMH), or kerosene as fuel. A list of some major launch vehicles (not an exhaustive list) is provided in Table 4.1.

Other than being used as the first stage rocket engine of launch vehicles, liquid propellant propulsion is commonly used in the upper stage as well. By using a series of precision valves and regulators, the operation of such a liquid propulsion system can be switched on and off repetitively. This not only allows the long duration firing but also offers the advantage of throttleability, which is not achievable in the solid propellant propulsion system.

The liquid propellant propulsion system is commonly employed as a spacecraft or satellite propulsion system. It has a rich heritage in both primary (e.g., orbit insertion, maneuver, and orbital plane change) and secondary propulsion functions (e.g., attitude control, station keeping, momentum wheel unloading, spin control, etc.). Currently, there are numerous commercially available liquid propellant propulsion

Table 4.1 Some selected launch vehicles from different countries.

	Country	Oxidizer	Fuel
Ariane 5	EU	Liquid oxygen	Liquid hydrogen
Atlas V	US	Liquid oxygen	Liquid hydrogen
Delta IV	US	Liquid oxygen	Liquid hydrogen
Falcon 9	US	Liquid oxygen	RP-1[a]
Long March 2, 3, 4	China	Dinitrogen tetroxide	Unsymmetrical dimethylhydrazine
Long March 5, 6, 7	China	Liquid oxygen	RP-1
Soyuz	Russia	Liquid oxygen	RP-1
Proton	Russia	Dinitrogen tetroxide	Unsymmetrical dimethylhydrazine
H-IIA, B	Japan	Liquid oxygen	Liquid hydrogen
GSLV MK II	India	Dinitrogen tetroxide	Unsymmetrical dimethylhydrazine
PSLV	India	Solid propellant: hydroxyl-terminated polybutadiene (HTPB)	

[a] *RP-1 (Rocket Propellant-1) is a refined form of kerosene used as rocket fuel.*

systems developed by companies like Aerojet Rocketdyne, Moog, IHI Aerospace, and Orbital Propulsion Center at ArianeGroup, which provide a wide range of propulsion solutions to meet the mission requirements of satellites of different sizes. For example, since 1966 ArianeGroup has produced more than 100 units of 25 N hydrazine monopropellant thruster for several ambitious and demanding missions while over 500 units of smaller version 1 N hydrazine monopropellant thruster have also been successfully operated in various space missions.

4.1.1 Operating principles

In a typical liquid propellant propulsion system, the liquid propellants are transferred from the storage tank into the combustion chamber. During the high-pressure combustion process, the chemical energy is released and converted into internal energy in hot reaction gas products. A supersonic nozzle, with a converging section down to a minimum area, known as the throat, and followed by a conical or bell-shaped diverging section, is then used to expand and accelerate the hot gases to achieve high exhaust velocity.

Depending on the chemical properties of the propellant, a liquid propellant propulsion system can be further classified into bipropellant or monopropellant. For bipropellant system, a liquid oxidizer and a liquid fuel are mixed and combusted in the reaction chamber, as schematically shown in Fig. 4.1A. For a monopropellant system, a chemical liquid (either a mixture of oxidizer and fuel species or a single homogeneous chemical substance) is catalytically decomposed into hot gases, Fig. 4.1B.

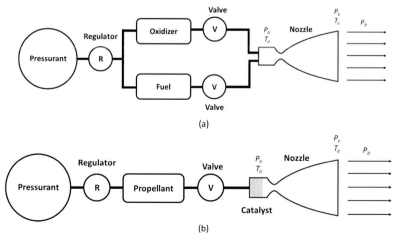

FIGURE 4.1

Schematic of a typical liquid propellant propulsion system: (A) bipropellant system; (B) monopropellant system.

High-pressure gas (pressurant) is commonly used to deliver the liquid propellant from a storage tank into the combustion chamber. This is straightforward for most small-scale liquid propellant thruster systems of low thrust level, usually in the range of milli-Newton (mN) to Newton (N). As the thrust level increases, the required static pressure in the tank to feed the liquid propellant at the desired mass flow rate has increased proportionally. As a result, a heavy tank with a thick wall is needed to withstand such high static pressure. This has increased the dry mass and adversely impacted the overall efficiency of the propulsion system as the mass ratio is reduced. Alternatively, a rather complicated but effective turbopump feeding system, which consists of a propellant pump, gear case, high-speed turbine, and other control valves, is used to transport a large amount of liquid propellant into the combustion chamber for those high-thrust liquid propulsion systems, such as the rocket engines in the launch vehicle.

4.2 Liquid propellants

The understanding of the characteristics and properties of different liquid propellants is crucial as the performance of any liquid propellant propulsion system is vastly dependent on the chosen propellant (or combination). By definition, liquid propellant refers to a variety of liquids that can undergo chemical and thermodynamic processes, which produce hot gaseous products. They can be categorized into three groups as follows:

1. Oxidizer. Some common examples are liquid oxygen, dinitrogen tetroxide, hydrogen peroxide, and nitric acid
2. Fuel. Some common examples are liquid hydrogen, kerosene, UDMH, and monomethylhydrazine (MMH)
3. Chemical compound (a mixture of fuel and oxidizer ingredients or a single homogeneous chemical substance), which undergoes self-decomposition in the presence of a catalyst. Some common examples are hydrazine and hydrogen peroxide.

4.2.1 Performance of propellant

From Newton's law of motion, the thrust produced by a liquid propellant propulsion system is given by

$$F = \dot{m} v_e \tag{4.1}$$

where $\dot{m}$ is the rate of propellant mass flow and v_e is the effective exhaust velocity, which can be related to the characteristic velocity by

$$v_e = C_F c^* \tag{4.2}$$

where C_F is the thrust coefficient, which is dependent on the nozzle properties while the characteristic velocity, c^*, is defined as

$$c^* = \sqrt{\frac{RT_0}{\gamma M}\left(\frac{\gamma+1}{2}\right)^{\frac{\gamma+1}{\gamma-1}}} \qquad (4.3)$$

From Eq. (4.3), it is evidenced that one of the key performance parameters for rocket propulsion, characteristic velocity, c^*, is a function of molecular mass (M) and the ratio of specific heats of the exhaust gas (γ), and temperature at the combustion chamber (or combustion flame temperature of the liquid propellant, T_0). All of them are closely related to the combustion of the chosen liquid propellant. The composition of combustion by-products determines the M and γ values. The more chemical energy released during the combustion process, the higher the flame temperature. It is noteworthy that the characteristic velocity is independent of nozzle characteristics, and therefore a good parameter to compare the performance of different liquid propellant propulsion systems. The flame temperature and the associated characteristic velocity for a few common combinations of propellant are shown in Table 4.2.

Apparently, the combustion of fluorine and hydrogen produces the highest flame temperature and yields the highest characteristic velocity. Nevertheless, this combination of bipropellant is only limited to experimental study due to the toxic and corrosive nature of fluorine, which is difficult to handle and store for any practical application in the liquid propellant propulsion system.

The molecular mass of the exhaust gas has a significant impact on the characteristic velocity as well. For the combination of liquid oxygen and kerosene (O_2/RP-1), the combustion flame temperature is 3701 K, which is higher than that of liquid oxygen and liquid hydrogen (O_2/H_2). The higher combustion flame temperature implies the more energetic combustion of a particular combination of propellants. However, the primary combustion by-product of carbon dioxide from the combustion of O_2/RP-1 propellant has increased the average molecular mass of the exhaust gas (M) notably if compared to the by-products of mostly H_2O, and unreacted H_2 or O_2 (depending on either fuel-rich or fuel-lean propellant mix) from the

Table 4.2 Performance comparison of a few selected combinations of propellant [1].

Oxidizer	Fuel	Flame temperature (K)	c^* (m/s)
O_2	H_2	3251	2386
F_2	H_2	4258	2530
O_2	RP1	3701	1783
N_2O_4	MMH	3398	1724
N_2O_4	N_2H_4 + UDMH	3369	1731

combustion of H_2/O_2 propellant. The net consequence is a higher characteristic velocity for O_2/H_2 combination.

Another merit to evaluate the performance of a given propellant is specific impulse, which is defined as

$$I_{sp} = \frac{F}{\dot{m}g} \tag{4.4}$$

where $\dot{m}$ is the consumption rate of the propellant.

It is a measure of propellant consumption efficiency. For a high-performance propellant, a smaller amount of propellant (i.e., smaller $\dot{m}$) is consumed to produce a similar magnitude of thrust. A higher specific impulse implies a better-performed propulsion system with low propellant consumption.

4.2.2 From bipropellant to monopropellant

In a bipropellant propulsion system, the oxidizer and fuel are stored separately and will only be mixed in the combustion chamber when the demand for thrust generation arises.

Theoretically, a liquid bipropellant system gives the highest specific impulse among all chemical propellant systems. The combination of liquid fluorine as oxidizer and liquid hydrogen as fuel, added with beryllium solid particles, has demonstrated a specific impulse as high as 480 s at sea level and 565 s in vacuum [2]. However, as previously mentioned, the liquid fluorine and beryllium particles are toxic, which limits them to a mere scientific investigation without further development into practical applications.

Historically, the nontoxic combination of liquid oxygen−hydrogen (LOX/LH2) is an ideal and efficient bipropellant candidate. It is commonly used in booster and upper stages of launch vehicles, where the propellant performance is of main interest and priority. When a more compact system is required, for instance in the booster stage, the combination of liquid oxygen−hydrocarbon bipropellant is a good alternative. Hydrocarbon fuels have been used extensively in many industries, for example, automotive, aviation, machineries, etc. A highly refined form of kerosene, designated as RP-1, is particularly suitable for use as hydrocarbon fuel in the rocket propulsion systems. Compared to liquid hydrogen, RP-1 has a higher density (0.81 g/mL vs. 0.71 g/mL for LH2), which allows a compact fuel tank design. In addition, RP-1 offers many attractive advantages as a fuel, such as ample supply with reasonably low cost, stable at standard atmosphere, and low explosion hazard, making it relatively easy to handle and store. Nevertheless, these two bipropellant combinations could not eliminate the need for cryogenic storage of liquid oxygen or liquid hydrogen.

To achieve a simplified and long-term propellant storage, the combination of liquid nitrogen tetroxide (N_2O_4) as oxidizer and hydrazine (and its derivatives) as fuel has been used in the past. If stored in appropriate conditions, for example, with sealed containers made of compatible material, the propellants remain stable indefinitely. Although the use of derivatives of hydrazine (i.e., UDMH and MMH) yields slightly lower performance in terms of specific impulse, they are more stable than pure hydrazine, even at higher storage temperature. For instance, UDMH has a lower freezing point at 215.9 K and a higher boiling point at 336.5 K. The ease in storage is indeed a huge advantage. In addition, the hypergolic characteristic of N_2O_4/UDMH (spontaneous ignition once mixed) has seen the earlier rocket engines use the combination of N_2O_4 and UDMH as propellants. Today, this combination of bipropellant is still widely used in various rocket engines, reaction control engines, or thrusters. A comparison of the performance of a few selected bipropellant propulsion systems is shown in Table 4.3.

A monopropellant propulsion system uses a chemical substance that contains both oxidizer and fuel components. It could be a mixture of several compounds, but most of the time, it is a single homogeneous chemical substance, for example, hydrogen peroxide, hydrazine, ethylene oxide, and nitromethane, to avoid further complication in storage and operation. While the energy content in monopropellants is comparatively lower than those in bipropellants, the simplicity in system design is attractive as only a unified propellant feeding and control system is required (instead of two separate sets of the system in bipropellant systems). Ideally, the monopropellant should be chemically stable with negligible degradation over the long-term storage, yet it can be decomposed easily into hot gases when thermally heated or catalytically activated. With these unique characteristics, a monopropellant propulsion system, when equipped with the appropriate fluid control system, is particularly suitable for attitude control, trajectory correction, and orbital maneuvers applications.

Of all monopropellants experimented with, hydrazine appears to be the most competent and established candidate. It is extensively used, even until today, in spacecraft or satellite propulsion systems. As an excellent storable monopropellant, it has a remarkable track record of 15 years of storage in a sealed tank [2]. When hydrazine is passed over a preheat catalyst bed, for example, iridium metals

Table 4.3 Performance comparison of bipropellant propulsion systems [1].

Propellant	Engine	I_{sp}, vacuum (s)	I_{sp}, sea level (s)
LOX/LH2	Vulcan 1	431	310
	RS-68	420	365
LOX/RP-1	RD-107	313	256
	Merlin	311	282
N_2O_4/UDMH	RD-253	316	285
	YF-21	289	259

supported by alumina bed, it is readily decomposed into ammonia, nitrogen, and hydrogen gas via the following simplified two-step reactions [2]:

$$3\,N_2H_4 \rightarrow 4\,NH_3 + N_2 \tag{R1}$$
$$4\,NH_3 + N_2H_4 \rightarrow 3\,N_2 + 8\,H_2 \tag{R2}$$

R1 is a highly exothermic reaction, which releases heat. In contrast, the decomposition of ammonia into nitrogen and hydrogen gases in R2 is an endothermic process that absorbs heat. Typically, a hydrazine based monopropellant system has a short ignition delay and yields a nominal specific impulse of 220 s [2].

4.2.3 From macroscale to microscale

Undoubtedly, the conventional liquid propellant propulsion system, be it bipropellant or monopropellant, is highly successful in macroscale operation, as evidenced by its enormous achievements in anything as large as a rocket engine to as small as a thruster. A similar success could not be simply replicated in microscale operation as there are many additional considerations before it becomes a norm to use liquid propellant propulsion in a nanosatellite.

LOX is an excellent and nontoxic oxidizer. However, the handling of LOX requires much attention. In the presence of ignition or spark source, LOX triggers and accelerates combustion. If LOX comes into contact with oils or organic materials at the same time, a detonation is possible. The cryogenic nature of LOX complicates the storage as the tank material must be made of material that can withstand cold temperatures. Furthermore, LOX evaporates rapidly, which does not favor long-term storage. Similarly, LH2 is highly flammable and explosive when mixed with air. As most of the metals are brittle at low temperatures, the associated cost to fabricate and insulate the piping system and storage tank for LOX and LH2 is deemed expensive for a nanosatellite mission.

Bipropellant system is characterized by its high performance as a result of high flame temperature combustion. From an operation point of view, the nozzle must be made of refractory metal alloys or ceramics of high melting temperature in addition to an appropriate cooling strategy, such as regenerative cooling, for long-duration firing. As the nanosatellite is inherently small in volume, dedicated thermal management is required to avoid the undesired overheating problem by shielding the excessive heat transfer from the high operating temperature of the bipropellant propulsion system to other subsystems. Again, this leads to an increase in design complexity and therefore inflates the development cost of a nanosatellite mission.

While hydrazine (and its derivative) monopropellants do not possess energy content as high as the bipropellants, the relatively low chamber temperature (approx. 800 K) allows the use of common engineering materials, such as most aluminum alloys and stainless steel, as structural materials for thruster and its components. Nevertheless, such an advantage shall be considered together with the toxicity and carcinogen nature of hydrazine. Prolonged exposure to the hydrazine vapors, through any form of inhalation, ingestion, or skin contact, should be within the OSHA 8-h personnel limit of 0.1 ppm. The rich heritage from previous missions has accumulated extensive protocols to handle it safely. For instance, personnel in Rocketdyne must wear self-contained atmospheric protective ensemble (SCAPE) and follows a 25-step procedure (as published in *Hydrazine Handling Manual*) when handling the hydrazine [3]. In addition, the well-established component level development in small monopropellant thrusters used in larger satellites has seen relatively less integration issue if it is implemented in nanosatellite mission as a primary propulsion system. However, the primary obstacle remains the adequate expertise in handling the toxic propellant as most of the present nanosatellite missions are carried out by institutions with limited experience in handling such toxic monopropellants.

4.2.4 Emerging of energetic ionic liquids as green propellant

Recognizing the increased concerns in cost and handling of the toxic hydrazine monopropellant, an initiative to look for a green alternative to replace hydrazine is gaining momentum. Such a green monopropellant is expected to benefit not only the conventional thruster systems for bigger spacecraft and satellites but also applicable in micropropulsion systems in the future. Three categories of the figure of merits, that is, toxicity, handling and transportation, and performance and material compatibility, have been established and proposed to assess the suitability of a green propellant [4].

Energetic ionic liquids (EILs), such as ammonium dinitramide (ADN) and hydroxylammonium nitrate (HAN), are widely regarded as promising alternatives to hydrazine. The level of toxicity of chemical propellant is commonly characterized using LD_{50}, which represents the amount of substance required (in mg) in causing the death of half of a tested population. Typically, the smaller is the value, the more lethal is the substance. To put it into perspective, the notorious and highly toxic Sarin gas has a LD_{50} value as low as 172 µg/kg and LD_{50} for water stands at 90,000 mg/kg. The LD_{50} values for hydrazine, ADN, and HAN are compared in Table 4.4. The LD_{50} for hydrazine is 59 mg/kg while it is 832 mg/kg and 325 mg/kg

Table 4.4 Comparison of key parameters between hydrazine and EILs.

	Hydrazine	ADN	HAN
Chemical structure	N_2H_4	$[NH_4]^+ [N(NO_2)_2]^-$	$[NH_3OH]^+ [NO_3]^-$
LD_{50} (mg/kg)	59	832	325
Density (g/cm^3)	1.021	1.81	1.85
Melting point (°C)	2	93	48
Specific impulse (s)	237	259 (FLP-106) 256 (LMP-103S)	257 (AF-M315E)

for ADN and HAN, respectively. Apparently, ADN and HAN have much lower toxicity, to an extent that SCAPE suits are not required when handling these green propellants [5], as shown in Fig. 4.2.

The low toxicity of EILs also simplifies the system and component-level design. According to MIL-STD-882E (Standard Practice for System Safety), the hazard severity of EILs is classified in the range of "critical" to "marginal," as compared to "catastrophic" for hydrazine. The lower hazardous risk means only a relatively simple two-seal strategy to inhibit the potential leakage [6] is required as opposed to the necessity to include isolation devices in the feedline system for "catastrophic" classification level.

In addition, the high solubility of these EILs in water is another advantage as ADN and HAN are stable and easy to store in an aqueous form. For example, the LMP-103 liquid monopropellant (60%−65% ADN, 15%−20% Methanol, 3%−6% Ammonia, the remaining % is water) developed by EURENCO Bofors is classified as UN and US DOT 1.4S. The storage packing of LMP-103 in the form of a polyethylene jug in a wooden box filled with adequate absorbent is sufficient for transportation by commercial passenger aircraft [7].

Both ADN and HAN have many desirable properties for their implementation as the next-generation liquid propellant for micropropulsion systems. In terms of performance, ADN- and HAN-based propellant blends have demonstrated approximately 6% higher specific impulse over hydrazine. The higher density of EILs also translates into (1) more green propellants packed into the same tank volume, or (2) smaller tank volume if the propellant mass is fixed as a constant, which subsequently results in an improvement in impulse density, as high as 30% when compared to a hydrazine-based propulsion system.

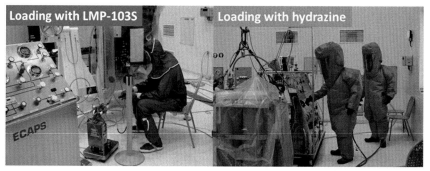

FIGURE 4.2

Comparison of safety suits worn by personnel when handling EILs (left) and hydrazine (right) propellants.

Reproduced with permission from ECAPS AB.

4.3 State-of-the-art liquid propellant microthruster
4.3.1 Hydrazine thrusters

Traditionally, 1 N monopropellant hydrazine thruster has been widely used for attitude control of bigger spacecraft and satellites and therefore possesses extensive
flight heritage in numerous missions. A few examples are given in Table 4.5. It
may not be proper to integrate these hydrazine thrusters directly into the nanosatellite buses. Nevertheless, it is expected such a thruster with a 1 N thrust level could be
used as a primary propulsion system for nanosatellites with some modifications.
This is probable as the hydrazine thruster system was frequently modified using
the available components to meet different mission requirements in the past.

In addition, the existing 1 N hydrazine thruster systems have desired properties
or features that promise a reasonable lead time to transform into a nanosatellite
thruster. For example, all of these thrusters were made with materials that are
compatible with hydrazine. This eliminates the need to search for suitable structural
material for the thruster body. The thruster dry mass is reasonable, for example,
290 g for ArianeGroup 1 N and 380 g for Moog MONARC-1 thrusters, for integration into nanosatellites. Electrical power rating of the thruster is within the power
budget of typical nanosatellites as well. For instance, the Aeroject Rocketdyne
MR-103 thruster consumes 6.32 W to preheat the catalyst bed and 8.25 W to operate
the valve. The thruster head assembly is approximately 60 mm in length, which is
fairly small and could be miniaturized further to integrate into the nanosatellite
bus. The IHI Aerospace MT-9 thruster (Fig. 4.3) can be constructed using ITAR-
free materials upon request by the customer, which facilitates the launching from
any country.

4.3.2 EILs-based green propellant thrusters
4.3.2.1 1 N HPGP propulsion system
As explained previously, the technical challenges and cost involved in handling the
toxic hydrazine propellant have limited the selection of hydrazine-based thruster for

Table 4.5 Examples of 1 N hydrazine thruster developed by different companies.

Manufacturer	Model	Thrust	Mass (kg)	Mission	Reference
Aerojet Rocketdyne	MR-103		0.33	Numerous big scale satellites and spacecraft	[8]
ArianeGroup	1 N		0.29		[9]
Moog	MONARC-1	1 N	0.38		[10]
IHI Aerospace	MT-9		-		[11]

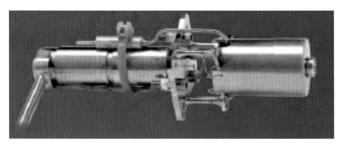

FIGURE 4.3

1 N MT-9 hydrazine thruster by IHI Aerospace.

Reproduced with permission from IHI Aerosace.

nanosatellite mission. Continuous efforts to develop an alternative green propellant-based propulsion system with higher performance and safer than the hydrazine system has seen the first breakthrough in 2010. The first ADN-based thrusters were successfully flight proven in orbit by ECAPS, a subsidiary of Swedish Space Corporation, in a mission called Prototype Research Instruments and Space Mission technology Advancement (PRISMA).

PRISMA mission consists of two satellites (MAIN and TARGET) to achieve two main mission objectives [6]:

(1) Demonstration of formation flying and rendezvous of two satellites in space using guidance, navigation, and control and sensor technology developed by other participating agencies.
(2) Demonstration of a high-performance green propellant (HPGP) propulsion system based on the green ADN-based monopropellant.

TARGET is a 50 kg satellite with no orbital control capability and coarsely stabilized with magnetic control. In contrast, MAIN is a 150 kg satellite, equipped with three different propulsion systems, and therefore highly maneuverable to perform a series of formation flying and rendezvous experiments around the TARGET satellite. The primary propulsion system of the MAIN satellite comprises six flight-proven 1 N hydrazine thrusters to provide the necessary ΔV capability to achieve the primary mission objective [12]. The two 1 N HPGP thrusters were planned to operate separately (from the hydrazine thrusters) as in-orbit demonstrators. However, once the HPGP thrusters were successfully demonstrated, they can operate concurrently with the hydrazine thrusters to provide additional ΔV capability to the mission and serve as redundancy when needed. These two systems provided a total ΔV of 170 m/s, which is more than double the mission requirements of PRISMA. The third propulsion system is another in-orbit demonstration of a microelectro-mechanical system (MEMS)-based cold gas micropropulsion system. A more detailed discussion about this system has been given in Chapter 3.

Early development works by ECAPS have confirmed ADN as a promising green monopropellant. Since its founding in 2000, ECAPS has tested several different

ADN-based propellant blends. In 2004, ECAPS has identified a propellant blend designated as LMP-103S with the composition of 63% ADN, 18.4% methanol, 4.6% ammonia, and 14% water, as the suitable ADN-based monopropellant for the 1 N HPGP thrusters of PRISMA. After extensive testing, LMP-103S shows superior performance in terms of combustion, storage stability, material compatibility, as well as sensitivity toward radiation. In one testing, the propellant blend showed no indication of performance degradation even after storing for more than 12 years [7]. Safety and hazard studies have confirmed LMP-103S as an insensitive substance, which can be loaded into a 5 L polyethylene jug and then packed in a wooden box with appropriate absorbent, Fig. 4.4, for shipping by air cargo. To date, LMP-103S has been shipped by air to launch sites or research institutions around the world. The exhaust gas species from the complete combustion of LMP-103S (50% H_2O, 23% N_2, 16% H_2, 6% CO, and 5% CO_2) are environmentally benign as well. The mild-to-moderate smell of ammonia from LMP-103S can be detected using ammonia sensors [7], which facilitates the triggering of early warning in case of any unwanted leakage in the system.

As LMP-103S demonstrates good compatibility with materials that are commonly used in the propulsion system, this allows the use of commercial off-the-shell (COTS) fluidic components with extensive flight heritage, especially those from the hydrazine-based system, with little or no modification during the development of the HPGP system. The propellant tank of the HPGP system has a capacity of 4.5 L (equivalent to 5.5 kg of LMP-103S), which uses a diaphragm to separate the helium pressurant gas and liquid monopropellant. The pressurant system is of blow-down type, where the feed pressure reduces from 2.2 MPa at beginning of

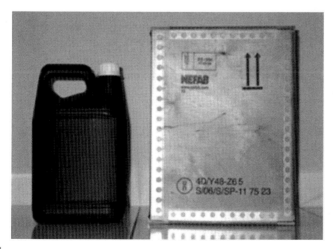

FIGURE 4.4

LMP-103S in transport package for air shipping.

Reproduced with permission from M. Persson, K. Anflo, P. Friedhoff, Flight heritage of ammonium dinitramide (ADN) based high performance green propulsion (HPGP) systems, Propellants, Explos. Pyrotech. 44 (9) (2019) 1073–1079.

life (BOL) to 0.55 MPa at end of life (EOL). As such, the thrust produced also varied proportionally from 1 N (BOL) to 0.25 N (EOL) [6]. A series of COTS fluid components, for example, service valves, isolation valves, pressure transducer, and propellant filter, were used to control the flow of propellant via 0.25-in. corrosion-resistance grade stainless steel tubing. A schematic of the fluid handling system for the HPGP system is shown in Fig. 4.5.

While there are many similarities in the fluid handling system design between hydrazine and HPGP systems, the design of thruster chamber assembly (TCA) is a vast difference. The thrust chamber and nozzle were made of iridium/rhenium alloy to withstand the high combustion temperature of LMP-103S, which could be as high as 1600°C during the steady-state firing as shown in Fig. 4.6. The conventional catalyst, S405, used in hydrazine thruster is not suitable for the HPGP system, and thus a dedicated high-temperature resistance catalyst has been developed by ECAPS.

The HPGP system cannot be cold-started. The catalyst bed must be preheated to a nominal temperature range of 340–360°C. A Thermacoat manufactured thermal control system, equipped with redundant heating elements and a K-type

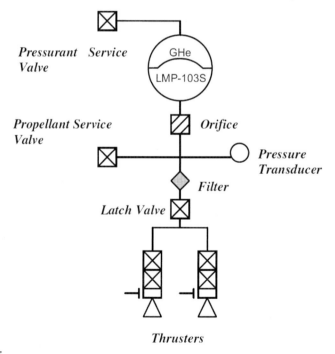

FIGURE 4.5

Hydraulic schematic of HPGP system.

Reproduced with permission from K. Anflo, R. Möllerberg, Flight demonstration of new thruster and green propellant technology on the PRISMA satellite, Acta Astronaut. 65 (9) (2009) 1238–1249.

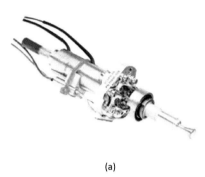
(a)

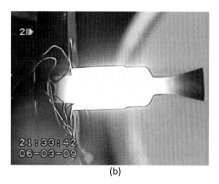
(b)

FIGURE 4.6

(A) 1 N HPGP thruster; (B) steady-state firing of 1 N HPGP thruster.

Reproduced with permission from K. Anflo, R. Möllerberg, Flight demonstration of new thruster and green propellant technology on the PRISMA satellite, Acta Astronaut. 65 (9) (2009) 1238–1249.

thermocouple, was used to control and monitor the operating temperature. Other than preheating, the heaters are only activated when the temperature drops below 350°C to save energy. On average, the power consumption is approximately 7.5 W for each thruster.

The PRISMA satellites were successfully launched on June 15, 2010, on the Dnepr-1 rocket, from Yasny/Dombarovsky launch site in Russia, into a 700-km sun-synchronous orbit. The mission was a huge success and could be deemed as a watershed for the development of EILs-based propulsion systems. There are many noteworthy achievements in this mission, both in space and on-ground. It has verified the previous claim that LMP-103S monopropellant (EILs in general) is safe and easy to handle on the ground as compared to hydrazine. The low hazard of LMP-103S has allowed the transportation by commercial aircraft and simplified prelaunch preparation as no SCAPE suit was required during propellant loading. The overall mission cost is notably lower than those that involve hydrazine propellant. Furthermore, PRISMA provided a meaningful direct performance comparison between hydrazine and HPGP propulsion systems as they are both rated at 1 N and developed using rather similar technology (except TCA). More importantly, they were operated using the same type of sensors (same satellite platform) and at the same space environment. Results from on-orbit steady-state firing, have revealed an 8% improvement in specific impulse for HPGP over hydrazine at BOL. The improvement reduced to 6% at EOL [7]. Over the 5-year mission, the HPGP also recorded an 18% improvement in ΔV per kg of propellant if compared to the hydrazine system.

After space qualified in the PRISMA mission, the 1 N HPGP thrusters were selected for implementation in another two small satellite missions. SkySat is a 24-satellite earth observation constellation that will ultimately provide submeter resolution imagery and video for commercial purposes. Each SkySat propulsion system has four 1 N HPGP thrusters, a flight heritage from PRISMA with an expanded three titanium propellant tanks to accommodate more LMP-103S propellants, Fig. 4.7.

FIGURE 4.7

Four 1 N HPGP system with three propellant tanks. Note: No SCAPE suits are required for handling personnel.

Reproduced with permission from M. Persson, K. Anflo, P. Friedhoff, Flight heritage of ammonium dinitramide (ADN) based high performance green propulsion (HPGP) systems, Propellants, Explos. Pyrotech. 44 (9) (2019) 1073–1079.

The propulsion system is able to deliver approximately 21 kN total impulse to satisfy the higher ΔV requirements for phase maintenance between satellites, drag compensation, as well as collision avoidance. Since 2016, 13 SkySat satellites with ECAPS' HPGP propulsion system have been launched and operated. The last batch of six Sky-Sat satellites with similar propulsion systems is expected to launch in 2020. STPSat-5, a small satellite of 115 kg developed for the department of defense Space Test Program, the United States, has been equipped with four 1 N HPGP thrusters as its propulsion system and was launched in December 2018 to LEO [13].

4.3.2.2 1 N green propellant reaction control system

In Japan, the research on HAN-based propellants has started as early as 2000 in a collaboration between the Institute of Space and Astronautical Science (ISAS) at Japan Aerospace Exploration Agency (JAXA) and Mitsubishi Heavy Industries (MHI) [14]. The propellant blend is designated as SHP163 composed of 73.6% HAN, 16.3% methanol, 3.9% ammonia, and 6.2% water. Compared to EILs-based propellant blend, adiabatic combustion of SHP163 produces the highest specific impulse at 276 s at a higher flame temperature of 2401 K, making it one of the most energetic green propellants [15].

SHP163 has undergone a series of safety evaluations and eventually was accepted as a green and safe liquid propellant. In 2011, a consortium consisting of Unmanned Space Experiment Free Flyer (USEF, formerly known as Japan Space Systems), ISAS, and MHI has jointly initiated the development of 1 N green propellant reaction control system (GPRCS) thruster. In 2005, the GPRCS system has been selected as one of the payloads in RAPid Innovative payload demonstration Satellite (RAPIS-1) mission (Fig. 4.8). RAPIS-1 mission is the first time that JAXA has

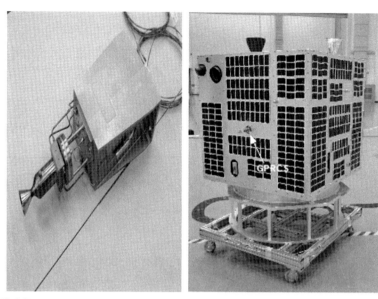

FIGURE 4.8

1 N GPRCS thruster and its installation on RAPIS-1 satellite.

Reproduce with permission from K. Hori, et al., HAN-based green propellant, SHP163 – its R&D and test in space, Propellants, Explos. Pyrotech. 44 (9) (2019) 1080–1083.

awarded the contract to a private company, Axelspace Corporation, to develop and manage the entire satellite. To reduce the mission complexity, it was decided to include the 1 N GPRCS system as part of the technology demonstration experiments, but not used as a main reaction control system for attitude and orbital control.

The development of 1 N GPRCS system employed a rather similar strategy as in PRISMA, where the flight-proven components inherited from hydrazine thrusters were used extensively, except the thruster body. It used a blow-down propellant tank with 0.9 MPa of pressure at EOL [16]. The loading of SHP163 propellant was done through the fill-and-drain valve without the need to wear a SCAPE suit at the launch site. This again verified the safe and green aspects of HAN-based green propellant in ground handling.

Under the long-term 2-step development plan by the consortium, a conventional iridium-based S405 catalyst was used in the STEP-1 to develop the 1 N GPRCS for RAPIS-1. The development of a heat-resistant high-performance catalyst will be carried out in STEP-2 with the aim to improve the overall performance thereafter [17]. Since S405 catalyst is not expected to withstand the high combustion temperature of SHP163, the firing process of the system must be properly tuned such that the ratio of propellant flow rate to catalyst mass is carefully controlled to avoid the overheating of the S405 catalyst beyond its servicing temperature. This comes with the expense of lower system performance as compared to the theoretical performance of SHP163. During the first test campaign, it was revealed that the catalyst weight reduced, and poisoning and structural deformation were spotted on the catalyst bed. However, the thruster performance was consistent after 2000 s of operation with the longest continuous operating time of 200 s [17]. In the last firing test before the launch, MHI has recorded an accumulated operation time of 5000 s in 2014. The completed 1 N GPRCS system has a dry mass of 5 kg and consumes a maximum of 16 W for preheating of catalyst.

As the largest satellite (200 kg) and the main payload for the Epsilon-4 rocket, the RAPIS-1 satellite was launched on January 18, 2019 from Uchinoura Space Centre. In the first in-orbit operation on February 18, 2019, the thruster was operated for 20 s. The success in operation was confirmed by the increased temperature measurement on the catalyst bed using a thermocouple. The thrust produced was validated from the increment in semimajor axis of the satellite orbit [15]. The mission is on course to achieve the target of 3000 s accumulated firing time and 10,000 pulses of firing in space.

4.3.2.3 1 N GR-1 thruster

A similar "green" revolution to replace hydrazine propellant has been taking place in the United States as well. Green Propellant Infusion Mission (GPIM) was initiated and managed by Marshall Space Flight Center at NASA, with Ball Aerospace and Technologies Corp as the prime contractor to perform a flight demonstration of HAN-based propulsion system in a Ball BCP-100 spacecraft bus [18]. The mission aims to showcase the performance improvement, low toxicity as well as safe handling features of a specialty HAN-based propellant, designated as AF-M315E, which were developed by the Air Force Research Laboratory [19]. Once the new green propulsion system has been space proven, it could be used as an effective and sustainable propulsion solution in the future NASA and commercial spaceflights.

The GPIM green propellant propulsion system has five 1 N GR-1 thrusters, Fig. 4.9A, built by Aerojet Rocketdyne. The thrusters employed AF-M315E as a monopropellant. The actual composition of AF-M315E remains classified. As AF-M315E has been proven as compatible with 6Al−4V titanium (a metal commonly used in hydrazine thruster), GR-1 thrusters used components inherited from the hydrazine system extensively. The overall GPIM system configuration consists of a diaphragm propellant tank, latch valve, filter, and service valve, which follows that of the flight-proven hydrazine system (TRL 9) closely, as shown in Fig. 4.9B. For example, it used the same ATK model 80581 hydrazine propellant tank with blow-down pressure in the range of 0.69−3.79 MPa [19,20]. The latch valve, pressure transducer, and filter are compatible with AF-M315E and were used without modification.

Considering the notably higher combustion temperature of AF-M315E propellant, the GR-1 thruster features an improved thermal isolation design via unique two-piece extended stand-off structures. Refractory alloys were only used specifically in high heat flux components, such as thrust chamber, nozzle, and thermal

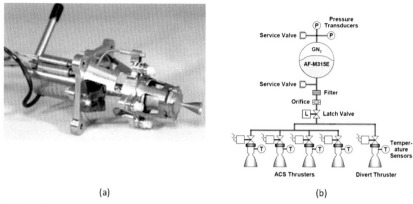

(a)   (b)

FIGURE 4.9

(A) 1 N GR-1 thruster; (B) system configuration.

Reproduced with permission from R.K. Masse, R. Spores, M. Allen, AF-M315E advanced green propulsion -
GPIM and beyond, in: AIAA Propulsion and Energy 2020 Forum, 2020.

isolation structure [20]. Other components exposed to lower heat flux could be made using the more cost-effective nickel alloys.

Aerojet Rocketdyne has replaced the low servicing temperature S405 catalyst used in hydrazine thruster with the LCH-240 catalyst, a 5% iridium catalyst coated on hafnium oxide ceramic support [19,20]. This newly patented catalyst has demonstrated exceptionally good resistance to the decomposition of AF-M315E propellant and thruster life extension over 15 times [20,21]. The new catalyst requires a higher nominal preheating temperature of 315°C, resulting in higher heating power consumption at 10 W. However, the increase in power consumption is partially offset by the very low freezing point of AF-M315E (at −80°C), which does not require much power to actively maintain the storage temperature as in the case of hydrazine propellant.

The GPIM satellite was launched on June 25, 2019, into a 722.3 × 711.1 km orbit by a SpaceX Falcon Heavy launch vehicle. After 4 days of commissioning activities, the propulsion system has started to operate and performed a series of functional testing. A Collision Avoidance maneuver was executed 2 months after launch, followed by other preplanned performance verification tests, such as impulse bit operations, space vehicle (SV) mass properties measurement, and thruster pulse count, as well as a demonstration of attitude control pointing, detumble, and momentum management.

A few months before the launch, the flight plan of the GPIM mission has been modified to perform multiple ΔV firings to lower the space vehicle altitude so that some useful scientific data could be collected at lower orbital altitudes [21]. Simultaneously, the ADCS performance could be monitored and the data collected could be used in low altitude drag and reentry modeling. To accommodate for the new flight plan, four perigee lowering burns were planned to reduce the altitude from

720 to 575 km, 300 km, and 180 km in stages. After executing the orbit lowering, final burns could be done to deplete the remaining propellant before the altitude of GPIM SV drops below 180 km and subsequently reentry within 2 weeks under the influence of atmospheric drag, which marks the end of the GPIM mission.

4.3.3 From small satellites into nanosatellites

The potential of EILs-based green propellant as an alternative to hydrazine propulsion systems has been demonstrated in the three small satellite platforms by different agencies, respectively. The technological know-how obtained from those missions paves a solid foundation for the further development of a nanosatellite-compatible green propellant micropropulsion system. The development is summarized in Table 4.6 and elaborated thereafter.

4.3.3.1 LituanicaSAT-2

LituanicaSAT-2 is a 3U CubeSat developed by Vilnius University. As part of the QB50 project, it is a technology demonstration mission. One of the mission objectives is to demonstrate a new high-performance and environmentally friendly chemical propulsion system in orbit. The experimental enabling propulsion system for small satellites (EPSS) prototype was developed by NanoAvionics, employed the LMP-103S ADN-based green propellant (space proven in PRISMA), and is capable to provide a steady thrust of 0.3 N, nominal impulse of 1 Ns, and ΔV up to 200 m/s. The peak power consumption of EPSS is 9 W and has a system dry mass of 0.6 kg [22,28]. The micropropulsion system could be used for multiple purposes, such as attitude control and drag compensation, and its high-thrust-to-weight ratio is powerful enough for Hohmann orbital transfer and collision avoidance maneuver by any 3U CubeSat.

The 3 kg LituanicaSAT-2 was launched on June 27, 2017, by the PSLV-C38 launch vehicle into a sun-synchronous near-circular orbit of 505 km altitude, making it the first nanosatellite operates with a chemical micropropulsion system.

4.3.3.2 M6P

The successful flight demonstration in LithuanicaSAT-2 has increased the TRL of EPSS to TRL 7, which is promising for further development into a commercial product. Based on the validated EPSS, NanoAvionics has upgraded the system with the implementation of modular design, where the mechanical and electrical components were preintegrated and functionally tested to facilitate the immediate integration with the nanosatellite bus. The volume of the tank can be scaled according to the need of the customer to meet mission requirements.

The improved system is designated as EPSS C1, which is 1U in size, Fig. 4.10 [29]. It can provide a thrust of 0.22 N at EOL and up to 1 N at BOL. The preheating power is 9.6 W and the dry mass of the system is 1 kg, and both are slightly higher than its predecessor EPSS [22,23].

Table 4.6 Summary of development in EILs based green propellant propulsion systems.

Propellant	Manufacturer	Thrust (N)	Mission	Satellite Mass	Year	Reference
ADN (LMP-103S)	Bradford ECAPS	1	PRISMA	150 kg	2010	[6]
			STPSAT-5	115 kg	2018	[13]
			SkySat constellation	120 kg	2016–20	[7]
HAN (SHP163)	USEF, ISAS, and MHI	1	RAPIS-1	200 kg	2019	[15]
HAN (AF-M315E)	Aerojet Rocketdyne	1	GPIM	180 kg	2019	[21]
ADN (LMP-103S)	NanoAvionics	0.3	LituanicaSAT-2	3 kg	2017	[22]
		0.1	M6P 1	6U[a]	2019	[23]
	Bradford ECAPS	0.1	ArgoMoon	14 kg	2021[b]	[24]
		1	ELSA-d	20 kg	2021[b]	[25]
		0.1	Lunar flashlight	14 kg	2021[b]	[26]
Water (LOX/LH2)	Tethers unlimited	1.2	Pathfinder technology Demonstration	11 kg	2021[b]	[27]

[a] No exact mass was disclosed in the literature. M6P 1 is a 6U CubeSat.
[b] Upcoming missions with expected launch date as published in the literature.

FIGURE 4.10

EPSS C1.

Reproduced from A.E.S. Nosseir, A. Cervone, A. Pasini, Modular impulsive green monopropellant propulsion system (MIMPS-G): for CubeSats in LEO and to the moon, Aerospace 8 (6) (2021).

EPSS C1 has been included in M6P, a 6U CubeSat developed by NanoAvionics using their new CubeSat platform. EPSS C1 has not only provided the typical propulsive capabilities as EPSS but is also used for deorbiting purposes at the end of the mission to meet the space debris mitigation requirements. The 6U CubeSat was launched on April 1, 2019, by the PSLB-QL launch vehicle. Upon completion of the mission, the EPSS will be able to achieve TRL 9.

4.3.3.3 ELSA-1d

Bradford Space has acquired ECAPS in 2017 and renamed it as Bradford ECAPS. The successful 1 N HPGP thruster will be featured in the upcoming End-of-Life Service by Astroscale (ELSA-d) mission. ELSA-d is the first project between Astroscale and SSTL to demonstrate some key technologies, such as rendezvous, docking, and proximity operations, for removing increasing space debris in orbit for sustainable use of the space environment [30].

ELSA-d comprises of "Chaser" small satellite (180 kg) and "Target" nanosatellite (20 kg). During launch, the two satellites are attached and will only be separated in-orbit. The mission plan includes capture maneuvers where the "Chaser" will locate the "Target" and perform a rendezvous and docking [25]. The flight-proven 1 N HPGP thruster by Bradford ECAPS will provide the necessary propulsive capabilities to the "Target" nanosatellite.

The ELSA-d satellites are scheduled to launch on March 2021 by Soyuz-2-1b Fregat M launch vehicle into a 550-km sun-synchronous orbit. Toward the end of the mission after approximately 6 months, the two satellites will be deorbited together.

4.3.3.4 ArgoMoon and Lunar Flashlight

Riding on the extensive experiences gathered from the 1 N HPGP thrusters in the PRISMA mission, Bradford ECAPS has miniaturized the thruster for lower thrust level operation. The resulting 100 mN HPGP Thruster, Fig. 4.11, is highly compact and uses nontoxic LMP-103S ADN-based propellant, making it particularly suitable for nanosatellite missions.

ArgoMoon is a 6U CubeSat for a deep space mission. It is selected by ESA to fly in NASA's Orion EM-1 (Exploration Mission-1) maiden flight. The mission will be the first CubeSat developed by a European country to leave the Earth orbit. In this historic mission, the ArgoMoon CubeSat will be the first satellite to deploy from the second stage of the space launch system (SLS) [31]. Using its dedicated hybrid micro propulsion system (MiPS), the 14 kg ArgoMoon will perform proximity maneuvering around the second stage of SLS, which allows the capture of detailed images of the second stage and records the historical moment of sending Orion into a lunar trajectory.

After achieving the first mission objective, ArgoMoon will move into a highly eccentric orbit with apogee close to Moon for a few more months to capture more images of the Earth—Moon system before the end of the mission. To fulfill the diverse propulsion requirements in attitude control and orbital maneuvering, the 100 mN HPGP thruster from Bradford ECAPS is combined with four 25 mN cold

FIGURE 4.11

100 mN HPGP thruster by Bradford ECAPS.

Reproduced with permission from ECAPS AB.

gas thruster from VACCO to form the hybrid MiPS for the ArgoMoon mission. The MiPS occupies a 1.3U volume. The 100 mN chemical thruster provides 56 m/s of ΔV and 783 Ns of total impulse while the 25 mN cold gas microthruster provides 72 Ns of total impulse and will be used primarily for attitude control [24].

Another 6U CubeSat onboard the Orion EM-1 will make it to the lunar orbit. The 14 kg Lunar Flashlight, managed by Jet Propulsion Laboratory and Marshall Space Flight Center, will become the first CubeSat to reach the moon and use 100 mN HPGP thruster (with LMP-103S propellant), signifying the first time that hydrazine is replaced with green propellant in planetary mission [26,32].

4.3.3.5 Pathfinder Technology Demonstration (PTD)

Tethers unlimited has developed an innovative micropropulsion system, called HYDROS (Fig. 4.12) [33], using water as a propellant. The improved HYDROS-C uses a microgravity compatible proton exchange membrane electrolyzer, to split the water into hydrogen and oxygen gases at an efficiency as high as 88%. The gases are stored in tanks separately and will only mix and combust in a bipropellant thruster on demand. The system consumes electrical power in the range of 5−25 W, depending on the rate of gas generation required. A maximum thrust of 1.2 N is achievable with a specific impulse as high as 310 s due to the highly efficient and clean combustion of H_2 and O_2 gases [27].

HYDROS-C adopts modular design approach, as the nonpressurized water allows the use of simple storage tank design of different shapes and dimensions to fit into CubeSat of different form factors (1−12U). Although HYDROS-C is

FIGURE 4.12

HYDROS thruster from Tethers Unlimited, Inc.

Reproduced with permission from K. Lemmer, Propulsion for CubeSats, Acta Astronaut. 134 (2017) 231−243.

categorized as a chemical micropropulsion system, it has zero stored energy before launching as the water propellant is not energetic material (as in other chemical propellants) before the electrolysis in-orbit. This results in an exceptionally high safety rating for HYDROS-C, which promises not only reduced safety risks to primary payload but also a high-performance micropropulsion system for orbital and large ΔV maneuvers.

To date, all the flight-proven chemical micropropulsion systems are based on EILs monopropellant. HYDROS-C will be integrated into a 6U CubeSat mission, Pathfinder Technology Demonstrator (PTD), which is slated to launch in early 2021. Upon successful mission, HYDROS-C will become the first bipropellant micropropulsion system operated in a nanosatellite platform.

4.3.4 Under development

This section discusses the liquid propellant microthrusters, which are yet to be flight demonstrated. Some of them achieved high TRL and waited for suitable flight opportunities, especially those from Aerojet Rocketdyne. A summary is given in Table 4.7.

4.3.4.1 MR-140 hydrazine thrusters

Aeroject Rocketdyne has started the development of hydrazine-based thruster for CubeSat as early as 2011, in an effort to increase the total ΔV capability, which was rather limited by the use of cold gas microthruster system. Leveraging on the rich experience in hydrazine thruster technology, the company has made further investments to miniaturize the components of the propulsion system. Using the MR-103 hydrazine thruster of numerous flight heritages as a baseline, MR-140 thruster

Table 4.7 Summary of liquid propellant microthruster under development.

Manufacturer	Propellant	Model	Thrust	Reference
Aerojet Rocketdyne	Hydrazine	MR-140	1 N	[34]
		MPS-120	1 N	[35]
	HAN (AF-M315E)	MPS-130		[35]
		GR-1A	1 N	[20]
		GR-M1	500 mN	[20]
IHI	HAN (HNP225)	Pinot-G	500 mN	[36,37]
BUSEK	HAN (AF-M315E)	BGT-X1	100 mN	[38]
		BGT-X5	500 mN	[39]
VACCO	HAN (AF-M315E) ADN (LMP-103S)	Green MiPS	100 mN	[40]
CU Aerospace	HTP/ethanol	MPUC	160 mN	[41]
Dawn Aerospace	N_2O/C_3H_6	PM200	500 mN	[42]

with a size reduction of approximately 4 times has been developed. MR-140 is producing a nominal thrust of 1 N. A CubeSat scale propulsion system, called CubeSat High-impulse Adaptable Monopropellant Propulsion System (CHAMPS), has employed four MR-140 thrusters. Carrying 360 g of hydrazine, CHAMPS produces approximately 800 Ns of total impulse [34].

4.3.4.2 MPS-120 hydrazine and MPS-130 green propellant thrusters

The hazard classification of hydrazine has limited the flight opportunity for CHAMPS, even though the system has completed the qualification testing. Aerojet Rocketdyne continued to develop the next generation modular propulsion systems (MPS) production lines to incorporate EILs-based propellant. New additive manufacturing technologies, such as electroforming (El-Form), selective laser melting (SLM), electron beam melting (EBM), and laser engineered net shaping, were introduced to streamline the manufacturing process and achieve cost saving. For example, the thrust chamber and nozzle are manufactured using the electroforming of Ir/Re alloy. The SLM was used to produce components, such as titanium bracket and Inconel stand-off. The propellant tank was manufactured using EBM. The new MPS lines yield the MPS-120 hydrazine thruster and MPS-130 green thruster using AF-M315E HAN-based propellant. It is noteworthy the two thrusters share the same design architecture and are rated at 1 N but use different catalyst beds. The modular approach allows MPS to expand easily from 1U (Fig. 4.13) into 2U configuration to carry more propellants and increases the total impulse

FIGURE 4.13

MPS-120/130 in 1U configuration.

Reproduced with permission from K. Lemmer, Propulsion for CubeSats, Acta Astronaut. 134 (2017) 231–243.

for more than two times, from 1130 Ns to 2720 as seen in MPS-130 thrusters (using AF-M315E propellant) [35].

4.3.4.3 GR-1A and GR-M1 thrusters

GR-1 thrusters in the successful GPIM mission came at a high unit production cost as the refractory material was used to produce thruster assembly and nozzle to withstand the high flame temperature of AF-M315E HAN-based propellant. To promote the adoption of the flight-proven GR-1 thrusters in future missions, in particular for nanosatellite, the unit cost must be reduced. Aerojet Rocketdyne has embarked on an initiative to reduce the cost by as much as 50% compared to the cost incurred in the GPIM mission [20]. A few possible approaches have been identified, such as to simplify the catalyst bed heater design to ease the assembly and installation process, substitute the refractory material in the thrust chamber structure with lower cost material where appropriate, use ceramic spacers at stand-off interface to reduce heat transfer, increased layering to reduce preheating power as much as 30%. CAD design of the upgraded GR-1A thruster is shown in Fig. 4.14. Other than the improvement achieved via design optimization, the next generation GR-1A thruster is expected to benefit from the improvement in the original ultrahigh temperature catalyst bed, now designated as LCH-240A, which has similar catalytic performance but at a lower production cost. The flight-ready GR-1A thruster is scheduled for qualification test by the middle of 2022.

In a parallel development, the GR-1 thruster has been miniaturized to a 500 mN thrust level. Designated as GR-M1, Fig. 4.15, it employs a similar thruster design approach, such as high-temperature resistant catalyst and refractory metals made

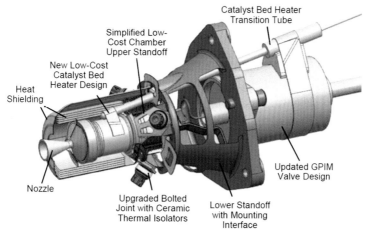

FIGURE 4.14

The upgrade GR-1A thruster.

Reproduced with permission from R.K. Masse, R. Spores, M. Allen, AF-M315E advanced green propulsion -
GPIM and beyond, in: AIAA Propulsion and Energy 2020 Forum, 2020.

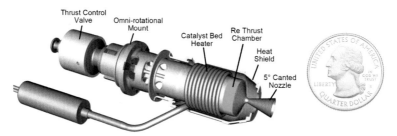

FIGURE 4.15

The miniaturized GR-M1 thruster.

Reproduced with permission from R.K. Masse, R. Spores, M. Allen, AF-M315E advanced green propulsion -
GPIM and beyond, in: AIAA Propulsion and Energy 2020 Forum, 2020.

thruster chamber, with the intended application in CubeSat. To overcome the challenging thermal management at a miniature scale, GR-M1 is operating on a 10 wt% water-diluted AF-M315E propellant. It is noteworthy that the dilution of AF-M315E propellant has lowered the specific impulse of 206 s (231 s in GR-1 thruster) [20]. The thruster design has implemented multilayer heat shields made of molybdenum-rhenium foil and encapsulated with Inconel shell to reduce the excessive heat transfer from the thruster chamber into the nanosatellite bus. The initial prototype is currently under hot fire testing to prepare for flight demonstration in the near future.

4.3.4.4 Pinot-G

IHI Aerospace has been working on HAN-based green propellants for more than 15 years in Japan. In contrast to the other research entities that strive to obtain the best performance from the EILs-based green propellants, IHI Aerospace is developing a new green propellant, which might not be the best performer but a good alternative of hydrazine thrusters for new players in the space industry, such as new space technology companies, research institutions, as well as universities.

To cater for the requirements of these emerging entities who might have smaller funding, IHI Aerospace introduced a cost-effective micropropulsion solution using a new HAN/HN/methanol/water green propellant, designated as HNP225 [36]. The actual composition of this new green propellant blend is classified.

From a safety point of view, HNP225 is safer and easier to handle as compared to other EILs-based propellants. According to United Nations Recommendations on the Transport of Dangerous Goods, both AF-M315E (UN Class 1.3C) and LMP-103S (UN Class 1.4S) are classified as Class 1 explosive substances. After a series of tests performed by the Nippon Kaiji Kentei Kyokai Physical and Chemical Analysis Center, the HNP225 propellant has been confirmed as a non-UN Class 1 explosive substance [37], which can be transported easily to end users.

According to data provided by IHI Aerospace (Table 4.8), the HNP225 has been modified from its predecessor (HNP209) such that the new propellant has a

Table 4.8 Comparison of HNP225 with hydrazine and other HAN-based green propellants.

	HNP225	Hydrazine	AF-M315E	LMP-103S
Density (g/cm^3)	1.00	1.17	1.46	1.24
Freezing point (°C)	1.5	<-10	<-80	<-6
Specific impulse, I_{sp} (s)	215	180	235	235
Combustion temperature (°C)	897	686	1353	1225

combustion temperature even lower than hydrazine. As such, the theoretical specific impulse of HNP225 is lower at 180 s as compared to hydrazine at 215 s. However, the density of HNP225 is higher than that of hydrazine. Combining these two factors, the new HNP225 is still offering a 5% improvement in terms of density impulse [36,37]. This implies that the HNP225 is equivalent to hydrazine but with much lower safety concerns, significantly simplifies the handling protocols and therefore the overall cost.

The lower combustion temperature of HNP225 eliminates the need for refractory materials in the manufacturing of thruster assembly, allowing the use of proven and more cost-effective materials, such as Inconel-625. To reduce the cost further, IHI Aerospace has manufactured a 0.5 N thruster using 3D printing technology, as well as using the conventional catalyst developed for the hydrazine system (Fig. 4.16A). Hot firing was conducted and the new thruster was operating nominally with stable combustion at a measured temperature of approximately 700°C, producing thrust in the range of 0.2−0.5 N at different inlet pressures. The successful testing will see four HNP225 based thrusters integrated into the Pinot-G micropropulsion system for application in nanosatellites in the future (Fig. 4.16B).

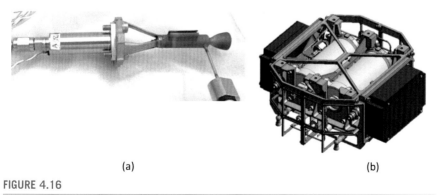

(a)   (b)

FIGURE 4.16

(A) 3D printed 0.5 N thruster by IHI Aerospace; (B) Pinot-G micropropulsion system.

Reproduced with permission from IHI Aerospace.

FIGURE 4.17

AMAC with BGT-X5 thruster.

Reproduced with permission from K. Lemmer, Propulsion for CubeSats, Acta Astronaut. 134 (2017) 231–243.

4.3.4.5 BGT-X1 and X5

Busek Co. Inc. is developing a green propellant micropropulsion system, specifically for CubeSats. The Advanced Monoprop Application for CubeSats (AMAC) is 1U in size and uses AF-M315E propellant, Fig. 4.17. The system uses HAN-compatible titanium extensively in the construction of a bellow propellant tank, thruster assembly, manifold, valves, and gaskets. AMAC can choose between two variants of thruster, 100 mN BGT-X1 and 500 mN BGT-X5. To accommodate the high combustion temperature of AF-M315E, a new nongranular and robust catalyst was developed by Busek. For stable combustion of AF-M315E propellant, the catalyst bed must be preheated to 430°C, consuming 20 W of power [38,39].

A distinctive feature of AMAC is the patented postlaunch pressurization system. Unpressurized at launch, the propellant tank will only be pressurized using gaseous CO_2 produced by heating the inert solid materials to 130°C in-orbit. This innovative solution also allows the repressurization of the tank after the pressure drops to a certain threshold until the depletion of solid materials, and thus maintaining a more stable thrust production throughout the mission.

The BGT-X5 prototype has been tested, a maximum thrust of 425 mN with 220 s of specific impulse was demonstrated. Integrated into a 1U volume, the system can produce precise continuous firing as well as impulse bits in the order of 0.05 N and is expected to deliver up to 146 m/s to a 4 kg CubeSat [38].

4.3.4.6 Green mono-propellant micropropulsion system

While contributing its cold gas micropropulsion system in the hybrid MiPS for ArgoMoon mission, VACCO has developed its own green propellant-based micropropulsion system. The VACCO's Green MiPS uses four 100 mN thrusters and is

3U in volume, carrying 2 kg of green propellant. The Green MiPS was developed using mainly titanium as structural materials and a high-performance catalyst bed, such that it is compatible with both AF-M315E and LMP-103S propellants.

Equipped with a dedicated controller system, each thruster in Green MiPS can operate independently for attitude control and ΔV maneuvers. Consuming a maximum electrical power of 15 W, Green MiPS is expected to deliver 3320 Ns of total impulse and ΔV of 237 m/s for a 14-kg CubeSat [40].

4.3.4.7 MPUC

CU Aerospace is developing a Monopropellant Propulsion Unit for CubeSats (MPUC) system using a new green and nontoxic propellant. The company is intended to search for such suitable monopropellant candidates among the easily sourced chemicals so that more could be benefited and not limited to those established space agencies and institutions that already possessed the technical know-how of EILs. In the preliminary search, it was shown that a chemical blend, designated as CMP-8, comprises 45% H_2O_2, 10.2% ethanol, and 44.8% H_2O, has a reasonable specific impulse of 180 s with a flame temperature of approximately 950°C [41]. The low flame temperature allows the use of common stainless steel instead of expensive refractory metals.

Using the in-house CUA catalyst formulation (CC-9) in the new catalyst bed, a thruster prototype (CMP-X) has been developed and a peak thrust of 174 mN was measured using a thrust stand with a continuous firing time exceeding 10 min. MPUC is currently at TRL 4, as shown in Fig. 4.18A is a CMP-X breadboard thruster. It is expected to achieve TRL 6 in 2022 for integration into 1U volume, Fig. 4.18B.

4.3.4.8 PM200

Hyperion Technologies and Dawn Aerospace have developed a bipropellant micro-propulsion system using nontoxic nitrous oxide and propene. Both fuel and oxidizer are industrial gases, that are available worldwide with no export restrictions, allowing the fueling by either Hyperion Technologies/Dawn Aerospace, or by the

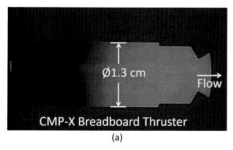
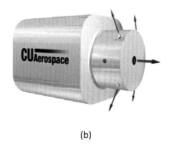

(a) (b)

FIGURE 4.18

(A) CMP-X breadboard thrust in a firing test; (B) a conceptual design of MPUC 1U.

Reproduced with permission from CUA.

FIGURE 4.19

PM200 bipropellant thruster by Dawn Aerospace.

Reproduced with permission from Hyperion Technologies.

customer. In addition, they are capable of self-pressurizing, which simplifies system design and improves safety rating.

In a typical 1U configuration, the PM200 weighs 1110 g (dry mass) and produces a nominal thrust of 500 mN at a specific impulse of 285 s (Fig. 4.19). It is expected to deliver 230 m/s of ΔV for a 3U CubeSat of 4 kg [42]. However, the use of additive manufacturing allows for compact designs and easy adaptability into other configurations offering optimized ΔV for a wider range of missions.

4.4 Challenges and future

4.4.1 Bipropellant micropropulsion system

The superior performance demonstrated by HYDROS-C and PM200 bipropellant-based micropropulsion systems has seen increasing research and development efforts in this area. In particular, the combinations of bipropellant that are nontoxic and easy to handle, which promises the ease of integration into nanosatellites, are much sought after.

A 0.4 N bipropellant micropropulsion is currently under investigation at the University of Miyazaki, Fig. 4.20. Nitrous oxide (N_2O) and dimethyl ether (DME) are chosen as the bipropellant with theoretical vacuum specific impulses as high as 290 s [43]. N_2O has been proven as an effective oxidizer in PM200. DME has a molecular

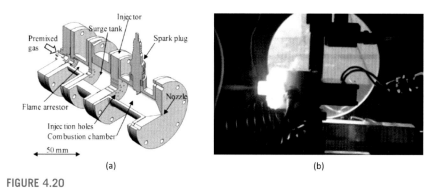

FIGURE 4.20

(A) 0.4 N N_2O/DME thruster prototype; (B) firing of thruster prototype in vacuum chamber.

Reproduced with permission from A. Kakami, A. Kuranaga, Y. Yano, Premixing-type liquefied gas bipropellant thruster using nitrous oxide/dimethyl ether, Aero. Sci. Technol. 94 (2019) 105351.

structure of CH_3-O-CH_3, the absence of the C—C bond leads to clean combustion with very low soot production. Both N_2O and DME can be stored in liquid form with desirable vapor pressure for self-pressurization, which simplifies the design of the propellant feeding system. The freezing temperature is considerably low at 293 and 131 K for N_2O and DME, respectively. Active antifreezing heaters are not required for the feed line and storage tank. In terms of safety, N_2O has LD_{50} value between 30,000 and 500,000 ppm while DME is rated at 164,000 ppm, both are much higher than 570 ppm for hydrazine. A 0.4-N thruster prototype has been tested, in which a C^* efficiency at 68.7% and a specific impulse of 135 s were reported.

An optimized mixing and efficient combustion would eventually release the full potential of bipropellant-based system in achieving a high level of thrust and specific impulse. From those early-stage studies, apparently, more investigations are still needed to improve the effectiveness in mixing of bipropellant. This is particularly challenging owing to the smaller scale of the combustion chamber that a micropropulsion system is operating. Identifying a new and innovative way of micromixing could potentially be a crucial enabling factor. Alternatively, hypergolic propellants could be the next object of investigation. Historically, hypergolic propellants have been used to eliminate the complicated mixing process as the propellants readily combust once they come into contact with each other. Nevertheless, the control of hypergolic reaction at the microscale remains a technical challenge to overcome.

4.4.2 Monopropellant micropropulsion system

From the literature covered in the previous section, it is evidenced that the EILs-based monopropellant micropropulsion system has achieved a good technological maturity and is ready to be widely implemented in various nanosatellite missions in the future.

The development trend in monopropellant micropropulsion systems suggests two diverse pathways. On one end, the EILs based systems are undergoing various optimization practices, such as improvement of catalyst durability at high combustion temperature, more energetic EILs blend by mixing with energy-rich fuel components, and cost effective yet high temperature-resistant structural material, all will boost the performance exceeding the hydrazine-based systems. Such development will benefit the high-value nanosatellite missions, which require a high-performance micropropulsion system to perform a wide range of demanding maneuvers in space.

On another end of the spectrum, more studies and development are expected in monopropellant micropropulsion systems based on the lower performance EILs blend, as pioneered in Pinot-G and GR-M1 systems. Concurrently, low toxicity monopropellant blend from easily sourced chemicals, as demonstrated in the MPUC system, will start to draw attention and further investigation. Compared to lower performance EILs, success in developing a liquid micropropulsion system based on such a new monopropellant blend could have far-reaching influence as a wider space community could access the monopropellant. The lower technological barrier offered by these systems allows wider adaptation of liquid propellant micropropulsion systems into nanosatellite missions by emerging and new stakeholders.

Until now, the lowest thrust level liquid propellant micropropulsion systems that are flight proven or flight ready for nanosatellite is 100 mN, for example, M6P1, BGT-X1, ArgoMoon, etc. When cold gas micropropulsion system is brought into the picture, a gap in micropropulsion solution between 25 and 100 mN with specific impulse above 100 s can be observed, as shown in Fig. 4.21.

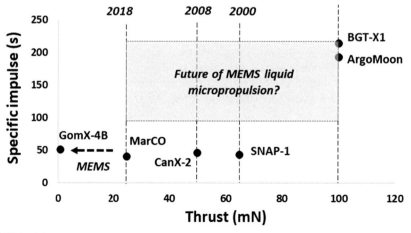

FIGURE 4.21

Opportunity gap for future MEMS liquid micropropulsion system.

The advancement in manufacturing technology has previously facilitated the reduction in thrust level in cold gas micropropulsion systems. In the 2000s, the micromachined cold gas microthruster in SNAP-1 and CANX-1 produced 65 and 50 mN, respectively. In 2018, the improved microfabrication technique has reduced the thrust level of cold gas microthruster of MarCO nanosatellites by approximately half to 25 mN. The use of MEMS microfabrication technique has further miniaturized the size of cold gas microthruster significantly to 10s of micrometer level. This leads to a thrust level as low as 1 mN in MEMS scale cold gas microthrusters in GomX-4b nanosatellite.

A similar MEMS initiative could see the thrust level of future liquid propellant micropropulsion system reduced beyond 100 mN and fill in the opportunity gap left by the cold gas micropropulsion system. Nevertheless, this remains a cautiously optimistic scenario until some technological hurdles are overcome in the future.

To date, some preliminary studies have been carried out to use silicon microfabrication technology to demonstrate the operation of bipropellant [44,45] and monopropellant [46–48] propulsion systems in MEMS scale. The system efficiency was rather low due to the enhanced heat transfer and reduced combustion efficiency at the microscale. Alternative materials, such as ceramic [49–52] and glass [48,53], have been used to replace the highly thermal conductive silicon as a structural material. The heat loss at microscale has diminished the preheating efficiency of catalyst bed for decomposition of monopropellants as higher electrical power consumption is required. Different decomposition techniques, such as electrolytic [49,54] and laser [55], have been tested, in search of a new and effective alternative.

Most of the studies are still in the proof-of-concept stage. From a system-level development perspective, although the size of MEMS-scale propulsion systems is much smaller, a successful and proven strategy to integrate the MEMS microthruster assembly with the supporting subsystems, such as the fluid handling and power systems, is notably absent now. While it is expected certain MEMS components inherited from the successful MEMS cold gas microthruster could be adopted, the high combustion temperature of liquid propellant requires additional considerations with probably some design modifications for elevated temperature operation.

References

[1] Turner, M.J.L., 2009. Rocket and Spacecraft Propulsion: Principles, Practices and New Developments, third ed. Springer.
[2] George, P., Sutton, O.B., 2001. Rocket Propulsion Elements, seventh ed. John Wiley & Sons.
[3] Hydrazine Handling Manual, 1961. Rocketdyne.
[4] Marshall, W.M., Deans, M.C., 2013. Recommended figures of merit for green monopropellants. In: 49th AIAA/ASME/SAE/ASEE Joint Propulsion Conference.
[5] Negri, M., 2015. Replacement of hydrazine: overview and first results of the H2020 project rheform. In: 6th European Conference for Aeronautics and Space Sciences (EUCASS).

[6] Anflo, K., Möllerberg, R., 2009. Flight demonstration of new thruster and green propellant technology on the PRISMA satellite. Acta Astronaut. 65 (9), 1238−1249.

[7] Persson, M., Anflo, K., Friedhoff, P., 2019. Flight heritage of ammonium dinitramide (ADN) based high performance green propulsion (HPGP) systems. Propellants, Explos. Pyrotech. 44 (9), 1073−1079.

[8] In-Space Propulsion Data Sheets, 2020. Aerojet Rocketdyne.

[9] Chemical Monopropellant Thruster Family Data Sheet, 2020. ArianeGroup.

[10] Monopropellant Thrusters Data Sheet, 2018. Moog.

[11] IHI Monopropellant Thrusters Data Sheet, IHI, 2020 Aerospace.

[12] Pokrupa, N., Anflo, K., Svensson, O., 2011. Spacecraft system level design with regards to incorporation of a new green propulsion system. In: 47th AIAA/ASME/SAE/ASEE Joint Propulsion Conference & Exhibit.

[13] Graham, C., et al., 2019. Balancing capability and cost on the STPSat-5 microsatellite mission. In: 33rd Annual AIAA/USU Conference on Small Satellites.

[14] Katsumi, T., et al., 2009. Combustion characteristics of a hydroxylammonium nitrate based liquid propellant. Combustion mechanism and application to thrusters. Combust. Explos. Shock Waves 45 (4), 442.

[15] Hori, K., et al., 2019. HAN-based green propellant, SHP163 − its R&D and test in space. Propellants, Explos. Pyrotech. 44 (9), 1080−1083.

[16] Tanaka, N., et al., 2011. The "greening" of spacecraft reaction control systems. Mitsubishi Heavy Ind. Tech. Rev. 48 (4), 44−50.

[17] Uramachi, H., et al., 2019. Green propulsion systems for satellites - development of thrusters and propulsion systems using low-toxicity propellants. Mitsubishi Heavy Ind. Tech. Rev. 56 (1), 1−7.

[18] Deininger, W., 2013. Implementation of the green propellant infusion mission (GPIM) on a Ball Aerospace BCP-100 spacecraft bus. In: 49th AIAA/ASME/SAE/ASEE Joint Propulsion Conference.

[19] Spores, R.A., 2015. GPIM AF-M315E propulsion system. In: 51st AIAA/SAE/ASEE Joint Propulsion Conference.

[20] Masse, R.K., Spores, R., Allen, M., 2020. AF-M315E advanced green propulsion - GPIM and beyond. In: AIAA Propulsion and Energy 2020 Forum.

[21] McLean, C.H., 2020. Green propellant infusion mission: Program construct, technology development, and mission results. In: AIAA Propulsion and Energy 2020 Forum.

[22] Propulsion System EPSS C1. Technical Overview by, 2020 NanoAvionics.

[23] Main Technical Specifications. CubeSat Propulsion System EPSS by, 2020 NanoAvionics.

[24] Data Sheet of ArgoMoon Propulsion System, 2017. VACCO Industries.

[25] Blackerby, C., et al., 2019. The ELSA-d end-of-life debris removal mission: preparing for launch. In: 70th IAC (International Astronautical Congress).

[26] Lunar Flashlight Propulsion System Data Sheet, 2017. VACCO Industries.

[27] James, K., et al., 2015. Performance characterization of the HYDROS™ water electrolysis thruster. In: 29th Annual AIAA/USU Conference on Small Satellites.

[28] Buzas, V., 2016. Expanding Smallsat Capabilities & Making Them Affordable. SatMagazine.

[29] Nosseir, A.E.S., Cervone, A., Pasini, A., 2021. Modular impulsive green monopropellant propulsion system (MIMPS-G): for CubeSats in LEO and to the moon. Aerospace 8 (6).

[30] Blackerby, C., et al., 2018. ELSA-d: an in-orbit end-of-life demonstration mission. In: 69th IAC (International Astronautical Congress).

[31] Tana, V.D., et al., 2018. ArgoMoon: challenges and design solutions for the development of A deep space small satellite. In: 69th IAC (International Astronautical Congress).

[32] Vinckier, Q., et al., 2018. System performance modeling of the lunar Flashlight CubeSat instrument. In: 49th Lunar and Planetary Science Conference.

[33] Lemmer, K., 2017. Propulsion for CubeSats. Acta Astronaut. 134, 231−243.

[34] Schmuland, D., Carpenter, C., Masse, R., 2012. Mission applications of the MRS-142 CubeSat high-impulse adaptable monopropellant propulsion system (CHAMPS). In: 48th AIAA/ASME/SAE/ASEE Joint Propulsion Conference & Exhibit.

[35] Masse, R.K., et al., 2013. CubeSat high-impulse adaptable modular propulsion system (CHAMPS) product line development status and mission applications. In: 49th AIAA/ASME/SAE/ASEE Joint Propulsion Conference.

[36] Igarashi, S., Matsuura, Y., 2017. Development status of a hydrazine alternative and low-cost thruster using HAN-HN based green propellant. In: 53rd AIAA/SAE/ASEE Joint Propulsion Conference.

[37] Igarashi, S., et al., 2019. Safe 0.5N green monopropellant thruster for small satellite propulsion systems. In: AIAA Propulsion and Energy 2019 Forum.

[38] Parker, K.I., 2016. State-of-the-Art for small satellite propulsion systems. In: Aerospace Systems Conference of the National Society of Black Engineers.

[39] BGT-X5 Green Monopropellant Thruster Data Sheet. BUSEK Space Propulsion and, 2021 Systems.

[40] Green Propulsion System (MiPS) Data Sheet, 2019. VACCO Industries.

[41] King, D., et al., 2016. Development of H_2O_2-based monopropellant propulsion unit for cubesat (MPUC). In: 63rd JANNAF Propulsion Meeting.

[42] PM200 Date Sheet. Dawn, 2021 Aerospace.

[43] Kakami, A., Kuranaga, A., Yano, Y., 2019. Premixing-type liquefied gas bipropellant thruster using nitrous oxide/dimethyl ether. Aero. Sci. Technol. 94, 105351.

[44] London, A.P., et al., 2001. Microfabrication of a high pressure bipropellant rocket engine. Sensor Actuator Phys. 92 (1), 351−357.

[45] Wu, M.-H., Lin, P.-S., 2010. Design, fabrication and characterization of a low-temperature co-fired ceramic gaseous bi-propellant microthruster. J. Micromech. Microeng. 20 (8), 085026.

[46] Hitt, D.L., Zakrzwski, C.M., Thomas, M.A., 2001. MEMS-based satellite micropropulsion via catalyzed hydrogen peroxide decomposition. Smart Mater. Struct. 10 (6), 1163−1175.

[47] Yuan, T., Li, A., 2009. Design and fabrication of a MEMS-based milli-Newton level hydrazine thruster. In: 45th AIAA/ASME/SAE/ASEE Joint Propulsion Conference & Exhibit.

[48] Huh, J., Kwon, S., 2014. Design, fabrication and thrust measurement of a micro liquid monopropellant thruster. J. Micromech. Microeng. 24 (10), 104001.

[49] Wu, M.-H., Yetter, R.A., 2009. A novel electrolytic ignition monopropellant microthruster based on low temperature co-fired ceramic tape technology. Lab Chip 9 (7), 910−916.

[50] Khaji, Z., et al., 2017. Endurance and failure of an alumina-based monopropellant microthruster with integrated heater, catalytic bed and temperature sensors. J. Micromech. Microeng. 27 (5), 055011.

[51] Cheah, K.H., Low, K.-S., 2014. Fabrication and performance evaluation of a high temperature co-fired ceramic vaporizing liquid microthruster. J. Micromech. Microeng. 25 (1), 015013.

[52] Cheah, K.H., Chin, J.K., 2014. Fabrication of embedded microstructures via lamination of thick gel-casted ceramic layers. Int. J. Appl. Ceram. Technol. 11 (2), 384–393.

[53] Huh, J., Seo, D., Kwon, S., 2017. Fabrication of a liquid monopropellant microthruster with built-in regenerative micro-cooling channels. Sensor Actuator Phys. 263, 332–340.

[54] Chai, W.S., et al., 2019. Calorimetric study on electrolytic decomposition of hydroxy-lammonium nitrate (HAN) ternary mixtures. Acta Astronaut. 162, 66–71.

[55] Alfano, A.J., Mills, J.D., Vaghjiani, G.L., 2009. Resonant laser ignition study of HAN-HEHN propellant mixture. Combust. Sci. Technol. 181 (6), 902–913.

Electric micropropulsions

Electrothermal microthruster

Angelo Cervone, PhD [1], Dadui Cordeiro Guerrieri, PhD [2],
Marsil de Athayde Costa e Silva, PhD [2], Fiona Leverone, PhD [2]

[1]*Assistant Professor, Aerospace Engineering Faculty, Delft University of Technology, Delft, the Netherlands;* [2]*Aerospace Engineering Faculty, Delft University of Technology, Delft, the Netherlands*

5.1 Historical background and principle of operation

Electrothermal propulsion can be seen as an intermediate concept between electrical and chemical propulsion. As schematically shown in Fig. 5.1, the propellant is heated electrically and accelerated thermodynamically, typically in a convergent—divergent nozzle. Propellant heating typically happens by means of a resistance (resistojet) or an electrical discharge (arcjet). For the extremely miniaturized applications that will be discussed in this chapter, resistojets are by far the most commonly used form of electrothermal thrusters.

In principle, any propellant can be used, stored in any phase (liquid, solid, or gaseous); in practice, however, liquid propellants are the most widely used. An alternative to liquid propellants is the so-called *warm gas* thrusters, which are basically cold gas systems allowing for additional (usually limited) heating of the gaseous propellant before being accelerated in the nozzle.

In terms of components and operational characteristics, as schematically illustrated in Fig. 5.1, electrothermal thrusters (and in particular resistojets) are very similar to cold gas thrusters: the propellant is stored in a tank, pressurized (when in the liquid phase) by a pressurant gas, and injected in the heating chamber by opening a thrust valve. A pressure regulator is usually not included, especially in miniaturized versions of this propulsion concept, meaning that it is typically operated blow-down and the pressure (and thrust) provided by the system is decreasing over its lifetime. The specific impulse, although higher than cold gas systems due to the higher temperature of the propellant at the nozzle inlet, is still limited due to the limitations in the available heating power and the maximum achievable temperatures.

Electrothermal thrusters have been employed so far in several space applications, since the 60s of the last century when a resistojet concept is reported to have been used on board of the *Vela* satellites [1]. In a detailed conference paper dating back to 1968, Mickelsen and Isley reported the state of art of resistojet thrusters at that time [2]. Particular emphasis is given in that paper to the TRW single-nozzle resistojet successfully flown in the *Vela III* satellite in 1965. This thruster, operating with gaseous

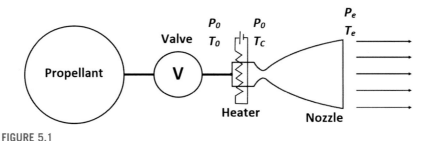

FIGURE 5.1

Schematic of a typical electrothermal propulsion system.

nitrogen, was reported to offer a thrust of approximately 187 mN at a specific impulse of 123 s and an input power of 42 W. A multinozzle version of this resistojet was then successfully flown on the *Advanced Vela* satellite in 1967, with a slightly improved specific impulse of 132 s and a thrust level of 89 mN per nozzle, each requiring a power of 17 W. A different type of resistojet, using liquid ammonia as a propellant, was developed by General Electric and flown on an unspecified NLR spacecraft in 1967. The thrust offered by each nozzle was 44 mN, requiring an input power of 3.2 W. The advantages of using a liquid propellant came at the price of a slight reduction in specific impulse, which was 100 s. Other more advanced electrothermal thruster concepts reported to be under investigation in the 60s included: using miniaturized needle-shaped devices to accelerate and heat the propellant; providing additional heating to the propellant in its supersonic stream in the nozzle; using a radioisotope thermal source to heat the propellant, instead of electrical resistance (the so-called "radioisojet"); or even, in manned spacecraft, using biowaste propellant to feed small resistojet thrusters for drag compensation or attitude control [2].

Despite this large amount of research on the resistojet concept, however, it never became a particularly popular option for use in spacecraft and satellites. Some applications of resistojets in commercial satellites are reported during the 80s of the last century, such as in the first satellites of the INTELSAT-V program. At the present date, Aerojet Rocketdyne offers on the market several electrothermal propulsion options, including both resistojets (MR-501) and arcjets (MR-509, MR-510, MR-512), which have been flown in a number of GEO satellites and in all the Iridium spacecraft. All these options, however, use hydrazine as propellant [3].

The research on miniaturized electrothermal thrusters, on the other hand, is still in its infancy and has not led yet to a significant number of flight-qualified concepts. However, as will be explained in detail in the following sections of this chapter, several universities, companies, and research institutions are conducting extensive research on different types of microresistojet concepts and ideas. The intrinsic simplicity of this propulsion concept, its adaptability to various possible propellants (solid, liquid, or gaseous), and its high thrust-to-power ratio when compared to other electric micropropulsion concepts make it a valid alternative for use in small and very small satellites, especially for the range of thrust in the order of 1−100 mN, which is at the margins of the typically available thrust level of both conventional miniaturized electric and chemical propulsion.

5.2 Current state of the art of electrothermal micropropulsion

A very good review of the existing electrothermal propulsion concepts is provided by Lemmers [4], to which part of the state of the art presented in this section refers. Other, less conventional options presented in the second part of this section are mainly based on the research performed by the authors of this chapter at the Delft University of Technology. It has to be noted, however, that most of the concepts presented in this section are still at an early stage of development and/or still require flight qualification in orbit. There is therefore still ample margin of development for their performance, effectiveness, and range of potential applications.

5.2.1 Conventional microresistojet thrusters

By "conventional" microthruster concepts, we refer here to either the so-called "warm gas" ones, in which a gaseous propellant is heated up (usually in a moderate way) before expelling it in the nozzle, or the vaporizing liquid microresistojets (VLM), where a liquid propellant is heated until vaporization and subsequently expelled in the nozzle.

One of the first warm gas MEMS propulsion modules demonstrated in orbit on a small satellite is the 3U module developed by Nanospace/GOMSpace [5]. This system based on Butane as a propellant, offered a total impulse of 40 Ns and a thrust resolution of 10 µN at a power consumption of 2 W, with a wet mass of 350 g and a size of $10 \times 10 \times 5$ cm^3 (corresponding to half a CubeSat unit). It can be used either as a fully cold gas system, without any heating of the propellant, or in warm gas mode, with slight propellant heating and consequent performance improvement. The full module includes four thrusters each with a thrust of 1 mN, at an operating pressure of 2−5 bar and closed-loop thrust control capabilities ensured by an embedded proportional flow control microvalve. It was demonstrated on the TW-1 satellite constellation, where it was reported to have been successfully used for along-track formation flight, allowing for fine-tuning of the orbital altitude with a control window of 0.5 km.

The CubeSat High Impulse Propulsion System (CHIPS) is a microresistojet system developed by CU Aerospace and VACCO [6,7]. The initial version of this system [6] was making use of the refrigerant gas R134a as propellant, with a specific impulse of 82 s, a thrust level of 30 mN, and a required power of 30 W. A successive version of the system [7], was reported to use R236fa as propellant, with a slightly decreased specific impulse of 68 s, at a thrust level of 20 mN and required power of 25 W. Although its performance is slightly worse than R134a, this alternative propellant (R236fa) was preferred because it allows for lower tank pressure and is compatible with the International Space Station filters, which in turn allows for increased safety in case of deployment from the Station. A peculiar characteristic of this system is that it includes a primary central thruster and four smaller attitude control thrusters. Also, in this case, it is possible to use the system either as a cold gas or as a resistojet by providing power to a "superheater cartridge," a small diameter tube in which the propellant flows before reaching the nozzle. By heating the thin walls of this cartridge, it is possible to provide the propellant with the required amount of heating before accelerating it in the nozzle.

A similar type of microresistojet system, but using ammonia as propellant and therefore offering a significantly higher specific impulse, has been developed by Busek Inc. [8]. Also, in this case, the system offers an integrated set of thrusters: a primary one (2−10 mN thrust, with a specific impulse of 150 s) and several, smaller, attitude control ones (0.5 mN thrust, 80 s specific impulse). The rated required power of the system is reported in the range from 3 to 15 W. The system has been laboratory demonstrated also with other propellants (including the refrigerants R134a and R236fa) and is designed to fit into a 1U CubeSat volume. It is claimed to be flight-ready, although no flight demonstration of this system seems to be reported in the open literature.

An interesting self-pressurizing microresistojet option is the RAMPART propulsion system, developed at the University of Arkansas [9] and based on 3D printing manufacturing technologies. The system has a rated thrust level of 500 mN and is designed for use with R134a as propellant. Similar to the previous examples, also in this case the thruster can be used either in a cold gas configuration (with a rated specific impulse of 67 s) or in a warm gas one, where the specific impulse can be increased up to 90 s. In this system, the propellant tank features a MEMS membrane with porous microchannels, that implements the surface tension of the fluid to separate the propellant liquid and gaseous phases. In this way, it is possible to use the vapor pressure of the propellant as a means to pressurize it in the tank, without any need for an additional pressurant gas. The nozzle and its integrated heater, made of several layers of bonded silicon, are based on MEMS manufacturing too.

A particularly interesting water microresistojet system, called AQUARIUS, has been developed by the University of Tokyo for the Japanese deep-space 6U CubeSat EQUULEUS [10]. This system is basically of the VLM type, but includes as particularly innovative feature a prevaporization of the water propellant in a separate vaporization chamber, which allows for complete separation between the thrust chamber and the heating chamber. Although at the cost of increased complexity (as an additional set of valves needs to be installed between the vaporization chamber and the thruster itself), this allows for a significant simplification of the heating process, which does not happen anymore in "real time" during the thrusting maneuver itself, but can be performed in advance in the vaporization chamber, with the actual acceleration of the vaporized propellant happening only after the vaporization is completed. This allows, in turn, for removing most of the fluid dynamics complications coming from the presence of a highly dynamic two-phase flow in the thrust chamber (instabilities, flow oscillations, supersonic collapse of the microbubbles). In addition, a further design feature of this microresistojet system is the use of waste heat from telecommunication components of the satellite as a power source for the vaporization chamber, thus reducing the power need of the system and increasing its intrinsic efficiency.

The AQUARIUS system is based on water stored at ambient pressure and temperature and subsequently heated in the vaporization chamber to 393 K at a pressure of 4 kPa. The nominal thrust is 4 mN, with a rated specific impulse of 70 s and a required power of 20 W. A flight prototype of AQUARIUS has been successfully demonstrated in orbit on the dedicated 3U CubeSat AQT-D, deployed in 2019 from the International Space Station [11].

An interesting alternative application of this concept, currently under research at the University of Tokyo, is in a "dual" configuration where the same water propellant is shared by two different propulsion units: a resistojet based on the same design of AQUARUS and an ion thruster with a thrust level of 0.25 mN, a specific impulse of 415 s, and a required power of 45 W [12].

As mentioned, the main difference between AQUARIUS and other VLM systems currently under development is the separation between the vaporization and the acceleration processes. Most of the other VLM systems, to the contrary, are based on two-phase vaporizing flow in a heater immediately upstream of the nozzle. This heater can be either "internal" (suspended in the flow and, therefore, directly heating the propellant) or, more frequently, "external", meaning that it is used to raise the temperature of the whole channel in which the propellant flows and, through the walls of the channel, to heat and eventually vaporize the propellant.

An example of this type of VLM microresistojet is the single-channel thruster developed and manufactured at the Indian Institute of Technology [13,14], Fig. 5.2 (left), which underwent an extensive experimental campaign and demonstrated a thrust of 1 mN using a maximum heater power of 3.6 W with a nozzle throat area of 130×100 µm. This thruster is based on a MEMS silicon chip in which one layer includes the heating channel and the convergent-divergent nozzle, while the layer on the opposite side includes a microheater made of a boron-diffused meanderline resistor in single crystal silicon. Another example is the pulsed microresistojet developed by Tsinghua University [15], Fig. 5.2 (right), working with pulsed heating with an average power of 30 W and generating over 1 s of thrusting a total impulse of 0.2 µNs. This microresistojet MEMS chip is based on a

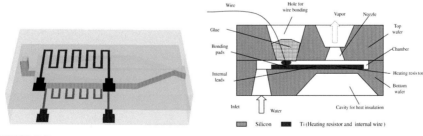

FIGURE 5.2

Left: 3D schematic view of the single-channel MEMS resistojet developed at the Indian Institute of Technology [14]. Right: schematic drawing of the MEMS resistojet developed by Tsinghua University [15].

combination of a silicon wafer with metallic Titanium of 200 nm thickness, used for the resistor and the internal wiring. Different from the previous concept, this one is based on an "out-plane" configuration (the propellant inlet and outlet are perpendicular to the silicon wafer) instead of an "in-plane" one (where the propellant inlet and outlet are along the silicon wafer axis). A thrust force ranging from 0.82 to 2.86 µN was experimentally measured for this microresistojet, at a pulse frequency of 30 Hz and a pulse width ranging, respectively, from 500 to 900 µs. Both thrusters shown in Fig. 5.2 use water as propellant.

However, the study and characterization of MEMS microresistojet thrusters date back to the work carried out around the end of the last century at the Jet Propulsion Laboratory of California Institute of Technology [16,17] and at the University of California in Davis [18]. The latter, see Fig. 5.3, proposed in particular two different designs of the microresistojet MEMS chip, an in-plane and an out-plane one. Water was used as propellant for the experimental characterization of the thrusters, achieving a maximum thrust level of approximately 0.19 mN with an input power of 6.7 W. As such, this thruster can be considered as the first laboratory demonstrated microresistojet concept based on MEMS manufacturing technology.

Finally, Delft University of technology is currently developing a MEMS microresistojet system for small satellites using liquid water as a propellant [19,20]. The thruster (Fig. 5.4) is based on a modular approach that allows to manufacture, in the same MEMS wafer, different thrusters based on different combinations of heating channels (either serpentine or diamond shaped) and nozzles. The thruster is heated by metallic molybdenum heaters and designed for a thrust in the range 0.6−1 mN and a specific impulse higher than 100 s, with a chamber pressure in the order of 5 bar. This VLM concept underwent several test campaigns that showed once again the extreme sensitivity of required input power to the chamber pressure, as well as the possibility of using the measured value of the heater resistance as an accurate measurement of the wall temperature in the heating chamber [21].

MEMS and their silicon-based philosophy are not the only options for VLM microresistojets. Some research has been conducted in the past on thrusters based on cofired ceramics, such as those at the National University of Singapore [22] and at Nanyang Technological University [23]. While the former was the first

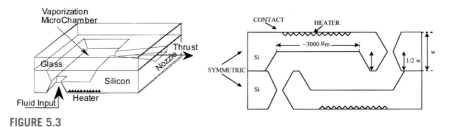

FIGURE 5.3

In-plane (left) and out-plane (right) microresistojet configurations proposed by the University of California in Davis [18].

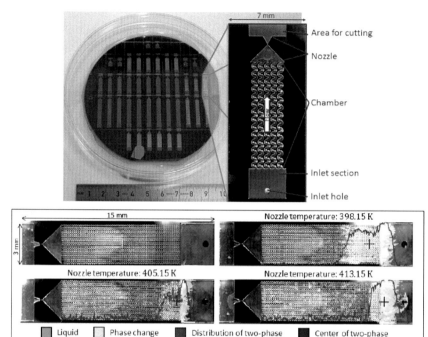

FIGURE 5.4

Top: MEMS wafer with different VLM thrusters developed at Delft University of Technology, and one specific thruster based on serpentine heating microchannels [19]. Bottom: vaporizing flow pattern in the thruster at different flow temperatures in the heating chamber [20].

research group to demonstrate the advantages of using this alternative material instead of silicon (in particular, its lower costs and relatively simple fabrication process), the latter further extended the concept by proposing a high-temperature design with a multilayer structure and a platinum heater (Fig. 5.5). The thruster was extensively tested, demonstrating 21% less power consumption compared to a VLM of similar characteristics based on silicon, but showing a particularly low specific impulse of just 31 s at a thrust level of 0.63 mN.

A summary of the VLM microresistojet concepts presented in this chapter is provided in Table 5.1. It includes some general info on each concept, as well as its main performance parameters (thrust, specific impulse, required power level) as reported in the open literature. This survey does not claim to be complete or fully exhaustive, but nevertheless it gives a very good overview of the type of VLM concepts developed so far and the performance level that can be expected from them.

The main technological challenge in VLM microresistojets is the presence of an intrinsically unstable two-phase flow in the heating chamber, which in turn leads to difficulties in controlling the thruster. Another challenge is associated with the need

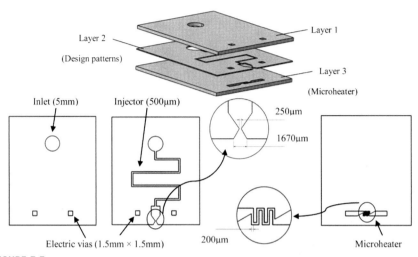

FIGURE 5.5

Schematic drawings of the high temperature cofired ceramic VLM proposed by Nanyang Technological University [23].

of using a micronozzle that, as already discussed in Chapter 3 of this book, suffers for significant flow losses when the Reynolds number in the nozzle falls below 1000, a typical situation for microresistojets delivering a thrust in the mN range.

For these reasons, research on the complex fluid dynamics inside the heating and thrust chamber of VLM microresistojets is of fundamental importance for a better design and understanding of these devices. An important contribution to this respect has been given by the research funded by the Taiwan National Applied Research Laboratories [24], which conducted an extensive analysis and experimentation on a VLM with a single channel heating chamber. They were the first to identify a clear relationship in VLM thrusters between mass flow rate, pressure, and heating power (see Section 5.4 for more details): for a given heating power provided to the thruster, the mass flow rate (and thus, the thrust) is a direct function of the pressure in the heating chamber. This means that, when pressure oscillations are present (e.g., due to instabilities caused by the two-phase flow), they generate in turn thrust oscillations, if the input power is kept constant, or as an alternative require to continuously control the input power as a function of the pressure, if the desired outcome is to keep stable thrust level. The flow visualization presented in Ref. [24] also led to the identification of four different flow patterns in the thruster, each with its own characteristics, depending on the mass flow rate, pressure, and channel geometry. This research was further extended by the Chinese Academy of Sciences [25], with an experimental campaign focused on a VLM with multiple parallel straight channels, which allowed to characterize several possible flow phenomena at the thruster microscale, such as presence of liquid droplets in the nozzle in case of insufficient heating, generation of vapor halos around small irregularities in the thruster,

Table 5.1 Overview of the VLM microresistojet concepts presented in this chapter (gray rows: options commercially available and/or flight qualified; white rows: options at prototype level).

Name	Developer (propellant)	Thrust [mN]	Isp [s]	Power [W]	Notes
Single-nozzle microresistojet	TRW (nitrogen)	187	123	42	Flown on the Vela III satellite
Multinozzle microresistojet	TRW (nitrogen)	89	132	17	Flown on the advanced vela satellite
Microresistojet	General electric (ammonia)	44	100	3.2	Flown on NLR spacecraft
MR-501	Aerojet rocketdyne (hydrazine)	370	303	493	Flown on BSAT-2 satellite
3U CubeSat propulsion module	NanoSpace/ GOMSpace (Butane)	1	n/a	2	Flown on the TW-1 satellite constellation. Also useable as cold gas.
CHIPS	CU Aerospace/ VACCO (R236fa)	20	68	30	Also useable as cold gas
Microresistojet system	Busek Inc. (ammonia)	0.5–20	80–150	3–15	Includes attitude control thrusters
RAMPART	Univ. Arkansas (R134a)	500	90	n/a	Manufacturing by 3D printing. Also useable as cold gas.
AQUARIUS	Univ. Tokyo (water)	4	70	20	Flown on the AQT-D CubeSat
Single-channel MEMS resistojet	Indian Inst. Technology (water)	1	n/a	3.6	Silicon chip with single heating channel
Pulsed MEMS resistojet	Tsinghua University (water)	0.8–2.9 $(\cdot 10^{-3})$	n/a	30	Designed for pulsed-mode operation
MEMS resistojet	Univ. California Davis (water)	0.19	n/a	6.7	First demonstrated MEMS resistojet ever
VLM MEMS resistojet	TU Delft (Water)	0.6–1	>100	n/a	Modular design, multiple parallel heating channels
Cofired ceramic VLM	Nanyang Technological University (Water)	0.63	31	n/a	Multilayer structure with platinum heater

gradual growth (in some cases degenerating into explosion) of large liquid droplets or even, when input power is excessively high, explosive boiling (Fig. 5.6). All these phenomena typically lead to significant pressure oscillations, which in turn make the controllability of the thruster more difficult.

The above considerations clarify the importance of modeling in an accurate way the flow inside VLM microresistojets, to promptly identify already in the early design phase any possible risks related to flow oscillations and instabilities. Particularly outstanding in this respect is the work conducted at the University of Salento [26,27], especially when looking at the relative simplicity of the models they proposed as compared to their predictive capabilities. The one-dimensional model presented in Ref. [26] is based on a coupled analysis of the steady-state flow in the heating chamber with the nonideal flow in the micronozzle, characterized by adding to the ideal rocket theory equations a semiempirical formula for the estimation of the discharge coefficient in presence of viscous losses. The results, obtained for the thrust and specific impulse of a VLM with multichannel heating chamber and parallel straight channels, showed agreement with the experiments with a relative error significantly lower than 10%.

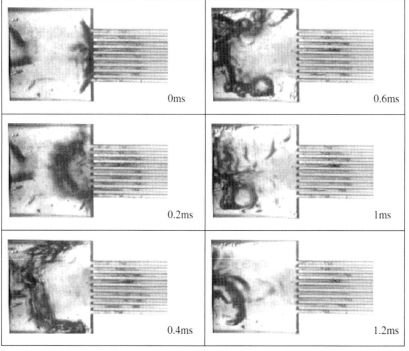

FIGURE 5.6

Optical visualization of explosive boiling in the inlet chamber of a VLM with parallel straight heating channels, caused by excessive input power to the thruster. The width of the inlet chamber is 2 mm, while the width of each microchannel is 80 μm [25].

5.2.2 A less conventional option: low-pressure microresistojet thrusters

As already mentioned, one of the main issues of miniaturized VLM microresistojets is the use of a micronozzle, which is intrinsically associated with significant flow losses when the Reynolds number becomes particularly low. One way to circumvent this specific issue is by operating the microresistojet at very low pressure, in such a way to work in the rarefied or transitional flow regime. This concept, known in the literature as free molecular microresistojet (FMMR) or more recently low-pressure microresistojet (LPM), allows for completely removing the conventional convergent-divergent nozzle and accelerating propellant particles from a low-pressure plenum only by means of collisions with the high-temperature walls of an expansion slot. Another advantage of this concept is that it allows for dual use, either as microresistojet or cold gas thruster, as it can also operate with no heating of the expansion slot walls.

The FMMR idea was first introduced and researched in the early 2000s by the Air Force Research Laboratory and the University of Southern California [28–31]. Their concept was initially based on a setting in which the heating surface is separated from the expansion slot (Fig. 5.7, left), subsequently improved into a geometry in which the walls of the expansion slots also act as heating surfaces (Fig. 5.7, right). Initial testing of this concept was performed with various candidate propellants (Helium, Nitrogen, Argon, Carbon Dioxide) and plenum pressure ranging from 50 to 300 Pa, demonstrating a thrust level in the range from 0.1 to 1.6 mN and a specific impulse in the range from 40 to 140 s with heater walls temperatures from 325 to 525 K. The same concept was successively demonstrated with water vapor, providing a thrust level of 0.13 mN at a specific impulse of 79 s [32].

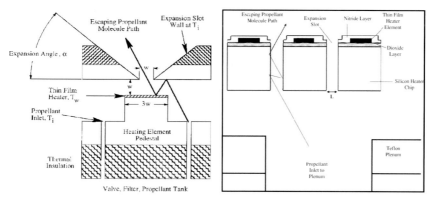

FIGURE 5.7

Left: the initial FMMR concept proposed by the Air Force Research Laboratory, with separated heating and expansion surfaces [29]. Right: the improved FMMR concept, using the expansion slot walls as heating surface [30].

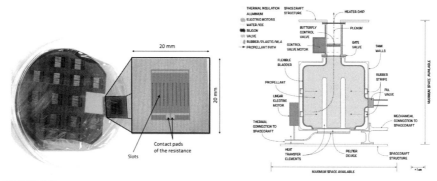

FIGURE 5.8

Left: MEMS wafer with different LPM thruster chips developed at Delft University of Technology, and one specific thruster chip based on rectangular expansion slots [33]. Right: sketch of a sublimating ice propulsion system based on the LPM concept [35].

The LPM concept has recently been brought into new life by the research conducted at Delft University of Technology [33–35], where an improved design of this type of microresistojet has been proposed, analyzed, and tested. This concept, intended to work with water vapor as propellant, is based on a MEMS thruster chip heated by molybdenum heaters, with different options, either rectangular or circular, for the geometry of the expansion slots (Fig. 5.8, left). The concept has been extensively tested and demonstrated a thrust in the range from 0.2 to 1.4 mN and a specific impulse in the range from 15 to 40 s, with the heater chip temperature varying from ambient to 149°C and the plenum pressure varying from 200 to 400 Pa.

This concept has also been proposed for use in combination with water stored in the solid state (ice) at a pressure below its triple point, to directly sublimate into vapor (Fig. 5.8, right). The idea is very promising, as it allows to combine the advantages of storing the propellant in the solid phase, with those coming from accelerating a rarefied gas in the thrust chamber. However, for this to become a feasible option, several challenges have to be solved, including in particular how to keep the propellant frozen from launch to actual operation in orbit.

5.2.3 An even less conventional option: solar thermal propulsion

Solar thermal propulsion (STP) is an alternative type of electrothermal propulsion in which a concentrator, such as a mirror or a lens, is used to focus sunlight either directly into the propellant or into a heat exchanger, or receiver (Fig. 5.9). Fiber optic cables can be used to transmit the concentrated solar energy from the concentrator to the receiver. Given the significant energy density available from solar thermal sources, propellant temperature can in principle be increased in this kind of system up to values in the order of 3000 K; however, the receiver can also act as energy storage system (based for example on phase change materials) to release this big amount of energy more gradually and allow for slightly lower propellant temperatures.

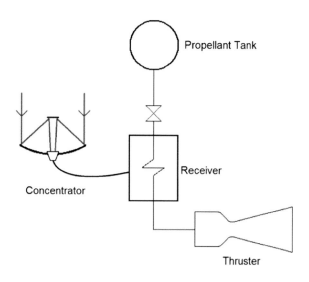

Propellant Tank

Receiver

Concentrator

Thruster

FIGURE 5.9

Schematic of a typical solar thermal propulsion system [36].

STP systems have recently been proposed also for smaller-scale applications, starting from the early 2000s, thanks to the technological advancements in the fields of fiber optics and thermal energy storage, as well as the possibility of employing them in bimodal integrated propulsion and power systems [36]. There are several current challenges to make these miniaturized STP systems actually available to space applications, related to both the concentrator and the receiver technology. For the concentrator, there is a need for more efficient, lightweight, and small optical devices with good optical efficiency, controllability, and deployment capability. For the receiver, the main problems still to be solved are related to insufficient sealing and lifetime of thermal cycling, as well as liquid containment issues. Finally, light-weight insulation materials are necessary to minimize the transfer of heat to other sensitive components of the spacecraft.

The available experimental results for miniaturized STP systems show that it is possible to achieve a thrust level in the order of 1 N at a chamber pressure of 2 bar, operating with a propellant temperature higher than 2000 K by using high-temperature materials such as graphite or rhenium [37,38].

5.3 Selection of propellant for electrothermal microthrusters

As previously indicated, one of the main advantages of electrothermal propulsion is that it allows for using virtually any propellant, as the only need is to provide heat to the propellant and eventually vaporize it. However, except for some specific cases (like the non-conventional idea of using sublimating ice in LPM thrusters, described in Section 5.2), it is preferable to use a propellant that can be stored in liquid state and, therefore, with sufficiently high density without requiring extreme pressurization as it would be the case for propellants stored in the gaseous state.

A detailed analysis of the possible propellant choices for microresistojets has been conducted by the authors of this chapter in Ref. [39]. A total of 95 different propellants were analyzed and prioritized, for both the VLM and LPM cases. As a first step, only propellants that can be stored as liquids (or solids) at ambient temperature, and pressure not higher than 10 bar, were taken into account. This left 63 candidate propellants, which were then ranked based on safety (flammability, instability, and health hazard), performance (specific impulse), and density. These criteria were given weights through a Pugh matrix process, based on the opinion expressed by experts on their relative importance. The final ranking is shown in Fig. 5.10, where not only the average scores are shown (based on the average weight of each criterion among all experts), but also their statistical deviations. More specifically, the middle red line of each box in the figure is the median score, the upper and lower borders of the boxes are the upper and lower quartile, the top and bottom lines are the maximum and minimum value, and the crosses are the outliers. It is clear from these results that nine fluids always score higher than the other ones, irrespectively on the specific expert opinion taken into account for the weighting of the criteria: acetone, ammonia, butane, cyclopropane, ethanol, isobutane, methanol, propene, and water.

These final nine propellants were then evaluated using, as baseline calculation case, the design and requirements of the VLM and LPM devices developed at the

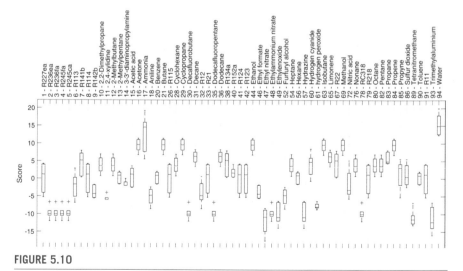

FIGURE 5.10

Global ranking of 63 candidate propellants for microresistojet thrusters, based on criteria weighted by means of a Pugh matrix process [39].

Delft University of Technology. Nevertheless, the results of this analysis (shown in Figs. 5.11 and 5.12) can be considered general and would apply to other designs as well, at least in terms of relative comparison between the propellants.

For each propellant, the theoretical specific impulse and the available Delta-V per unit volume of propellant were calculated, as functions of the required input power, for the full range of temperatures and pressures allowed by the requirements of the baseline thrusters. The performance parameters were calculated based on ideal simplified equations, similar to those presented in Section 5.4 of this chapter.

The results clearly show two best performing propellants: ammonia and water. The higher specific impulse of these propellants is mainly a consequence of their

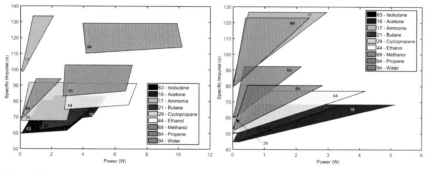

FIGURE 5.11

Specific impulse as a function of the input heating power for the nine best-ranked propellants, for the VLM (left) and LPM (right) case, in the range of pressures and temperatures allowed by the design and requirements of the devices developed at Delft University of Technology [39].

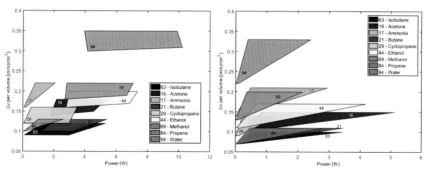

FIGURE 5.12

Delta-V per unit volume of propellant as a function of the input heating power for the nine best-ranked propellants, for the VLM (left) and LPM (right) case, in the range of pressures and temperatures allowed by the design and requirements of the devices developed at Delft University of Technology [39].

lower molecular mass, which allows for accelerating the lighter molecules of propellant at faster velocity with the same amount of energy. However, water suffers for a high value of the latent heat of vaporization, which in the VLM reflects a higher amount of required input power. This issue is not present in the case of the LPM concept, where the propellant is injected in the thruster already in the gaseous state and therefore does not need to be vaporized by the same heaters that are used to accelerate the molecules of propellant in the expansion slots. However, among these two propellants, water clearly shows a higher Delta-V per unit volume, in both the VLM and LPM cases, thanks to its higher liquid density. Therefore, especially for applications in which volume constraints are as important, if not even more important, than mass ones (as it often happens in small satellites and CubeSats), water is probably the best compromise between safety, performance, mass, and volume constraints.

A similar analysis has been conducted in Ref. [36] for STP systems, although for a more limited number of options (water, hydrogen, ammonia). The analysis considered in this case a realistic range of temperatures (1000–2500 K) and thrust levels (0.5 mN–2 N) for a typical STP system, with a nozzle expansion ratio of 100 and a chamber pressure of 2 bar.

The results are in this case less decisive, showing a completely different winner depending on whether the parameter of interest is the specific impulse or the Delta-V per unit volume (see Fig. 5.13). Given its very low molecular mass, hydrogen is clearly the most favorable option in terms of specific impulse, but at the same time shows very poor performance in terms of Delta-V per unit volume. More realistic alternatives in terms of Delta-V per unit volume are represented by ammonia and water, for which similar considerations as the VLM case apply: also, in this case, water seems to represent the best compromise choice between all possible selection criteria.

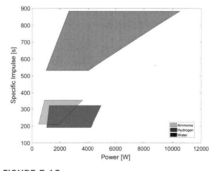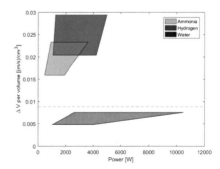

FIGURE 5.13

Specific impulse (left) and Delta-V per unit volume of propellant (right) as functions of the input heating power for three potential propellants, for the typical range of pressures and temperatures of a STP system [36].

5.4 **Theoretical analysis of conventional microresistojets**

Since VLM resistojets are typically based on accelerating the propellant in a conventional convergent-divergent nozzle, their performance can be analyzed in a simplified way by applying the ideal rocket theory equations. The assumptions on which these equations are based have already been introduced in Chapter 3 and can be summarized as follows:

- The fluid flowing in the nozzle is a perfect, calorically ideal gas of constant homogeneous chemical composition;
- Flow is steady, isentropic, mono-dimensional, with purely axial velocity;
- No friction or other external forces act on the gas flowing in the nozzle.

The nozzle is convergent−divergent, with an inlet section in which the flow is considered under stagnation conditions (negligible velocity), the throat section where the flow is sonic, and the exhaust section where the flow is typically highly supersonic.

Despite the large number of assumptions and simplifications on which it is based, the ideal rocket theory is still surprisingly accurate in evaluating the performance of the nozzle, with results that normally stay within 10%−15% from the actual values.

With reference to the schematic of Fig. 5.1, we assume that the liquid propellant enters the heating chamber at temperature T_0 and pressure p_0. The propellant is then heated up to higher temperature T_C (at which it is in the gaseous state), while it is assumed that no pressure drop takes place in the heating chamber, thus the pressure stays equal to p_0. The flow is then accelerated in the nozzle, being expelled from it at pressure p_e, temperature T_e, and jet velocity v_e.

Under the ideal rocket theory assumptions, it is possible to derive the following equation for the jet velocity:

$$v_e = \sqrt{\frac{2\gamma}{\gamma - 1} \cdot \frac{R_A}{M_W} \cdot T_C \cdot \left[1 - \left(\frac{p_e}{p_0} \right)^{\frac{\gamma - 1}{\gamma}} \right]} \qquad (5.1)$$

where R_A is the universal gas constant ($= 8314$ J/K × kmol), M_W and γ are the molecular mass and specific heat ratio of the gas flowing in the nozzle.

Eq. (5.1) shows that higher jet velocity can be achieved by selecting a propellant that allows for higher heating temperature and lower molecular mass of the gas flowing in the nozzle (which, in chemical engines, is a mix of the products of the chemical reaction in the chamber).

For the mass flow rate $\dot{m}$, the following equation holds:

$$\dot{m} = \frac{p_0 \cdot A^*}{\sqrt{\frac{R_A}{M_W} \cdot T_C}} \cdot \sqrt{\gamma \cdot \left(\frac{1 + \gamma}{2} \right)^{\frac{1 + \gamma}{1 - \gamma}}} \qquad (5.2)$$

where A^* is the nozzle throat area.

The thrust F_T produced by a propulsion system, as already mentioned in the previous chapters, can be calculated using the following equation:

$$F_T = \dot{m} \cdot v_e + (p_e - p_a) \cdot A_e = \dot{m} \cdot v_{eq} \tag{5.3}$$

where p_a is the external ambient pressure and A_e is the propellant exhaust area. Eq. (5.3) shows that the thrust is made of two different contributions: a "momentum term" (actual momentum exchange between propellant and spacecraft), and a "pressure term" (difference in pressure between the expelled propellant and the external ambient). To write the thrust equation in a more compact way, an *equivalent jet velocity* is usually defined, indicated by v_{eq} in Eq. (5.3), which accounts for both the momentum and pressure terms in the equation.

The specific impulse is defined as the ratio of the total impulse generated by the engine (thrust integrated over the burn time), to the total weight of propellant used to generate it. It is typically measured in seconds and gives a measure of the propellant consumption efficiency of the system: higher specific impulse means that higher total impulse is generated with the same propellant mass (or, alternatively, the same total impulse can be obtained by using less propellant). If the equivalent jet velocity is constant over time, the specific impulse I_{sp} can be simply written as follows:

$$I_{sp} = \frac{v_{eq}}{g_0} = \frac{F_T}{\dot{m}g_0} \tag{5.4}$$

where g_0 is always the gravitational acceleration on Earth at sea level ($= 9.81$ m/s^2), regardless of the place where the rocket or spacecraft is flying.

Finally, the Delta-V (ideal velocity change experienced by the spacecraft in which the propulsion system is installed, when a given mass of propellant has been expelled) is usually calculated by means of the *rocket equation*:

$$\Delta v = v_{eq} \cdot \ln\left(\frac{M_0}{M_0 - M_P}\right) \tag{5.5}$$

The rocket equation gives the velocity change of a spacecraft with initial mass M_0, when a mass M_P of propellant is expelled by its propulsion system with given equivalent jet velocity. However, this is only true under a number of assumptions: no external forces acting on the spacecraft (such as gravity or atmospheric drag); equivalent jet velocity constant over time; propellant expelled in a direction exactly opposite to the flight direction. When at least one of these assumptions is not met, the Delta-V calculated by means of the rocket equation is no longer the actual velocity change of the spacecraft; however, it is still a good indicator of the energy transferred by the propulsion system to the spacecraft, although only part of this energy actually contributes to increasing the actual kinetic energy of the spacecraft.

In a VLM system, Eq. (5.2) for the mass flow rate has to be combined with the following relationship that characterizes the vaporization and heating of the liquid propellant:

$$P_h = \dot{m} \cdot \left[c_{pL} \cdot (T_{boil} - T_0) + L_h + c_{pG} \cdot (T_C - T_{boil}) \right] \tag{5.6}$$

where P_h is the available heating power, T_{boil} is the propellant boiling temperature (which in turn is a function of the heating chamber pressure p_0), L_h is the latent heat of vaporization of the propellant. c_{pL} and c_{pG} are the constant pressure specific heat of respectively the liquid and gaseous propellant phase, both are usually functions of the temperature; however, in a simplified model, they can be considered constant and equal to their average value in the relevant range of temperatures. Combining Eqs. (5.2) and (5.6), it is possible to write the following relationship:

$$P_h = \frac{p_0 \cdot A^*}{\sqrt{\frac{R_A}{M_W} \cdot T_C}} \cdot \sqrt{\gamma \cdot \left(\frac{1+\gamma}{2}\right)^{\frac{1+\gamma}{1-\gamma}}} \cdot [c_{pL} \cdot (T_{boil} - T_0) + L_h + c_{pG} \cdot (T_C - T_{boil})] \qquad (5.7)$$

Eq. (5.7), despite having been obtained from an extremely simplified theory, clearly shows the direct relationship not only between heating power and chamber temperature (as expected), but also between heating power and propellant pressure. Although this result has been obtained starting from the ideal rocket theory equations and, therefore, with no flow losses involved, it can be considered of general validity also when losses are present. It confirms that, given a desired temperature at which the propellant has to be heated, the required heating power to achieve that temperature will be a function of the propellant pressure and, therefore, in a system without a pressure regulator it will vary over the lifetime of the system. Furthermore, pressure oscillations and flow instabilities like those typically observed in two-phase boiling flow, previously mentioned in Section 5.2, directly result in the need of controlling the heating power accordingly. This complicates the design of the control electronics and, at the same time, poses additional limitations on the achievable thrust and specific impulse levels.

The modeling equations for LPM microresistojets are significantly different, due to the rarefied flow conditions in this particular concept that make the ideal rocket theory (based on the assumption of continuum flow) not applicable. A simplified set of modeling equations is proposed in Ref. [40] and shortly summarized in the following.

The equations of this model are based on assuming thermodynamic equilibrium inside the LPM plenum and are obtained starting from a Maxwellian distribution for the thermal velocity of molecules in thermodynamic equilibrium. In this case, it is possible to show that the following equation applies for the mass flow rate:

$$\dot{m} = \alpha p_0 \sqrt{\frac{m_a}{2\pi k T_0}} \cdot A_e \qquad (5.8)$$

where m_a is the mass of a single molecule of gas, k is the Boltzmann constant, and A_e is the total exit area of all expansion slots. The parameter α in Eq. (5.8), called transmission coefficient, is crucial in this type of system. It is defined as the actual mass flow rate of molecules expelled from the expansion slots, divided by the ideal mass flow rate in the free-molecular limit; or, in other words, the ratio of the mass flow rate of molecules actually exiting the expansion slots, to the mass flow rate of molecules entering it. The two mass flow rates are not the same, as part of the molecules entering the expansion slots are bounced back and go back to the plenum without being expelled.

The transmission coefficient represents an important loss factor of LPM thrusters, as the molecules returning into the plenum, despite being partially energized in the expansion slot, are not expelled and therefore do not contribute to generating thrust. Semiempirical expressions for this coefficient can be found in Ref. [40], based on the geometry and shape of the expansion slot (in particular, for circular and rectangular ones). Typical values for expansion slots of constant area are in the range between 0.15 and 0.2, but it has been shown that they can be increased up to values higher than 0.7 with divergent expansion slots [34].

The jet velocity and exit pressure in a LPM system can then be written as follows:

$$v_e = \sqrt{\frac{\pi k T_{tr}}{2m_a}} \tag{5.9}$$

$$p_e = \frac{\alpha p_0}{\pi} \sqrt{\frac{T_{tr}}{T_0}} \tag{5.10}$$

where T_{tr} is the translational kinetic temperature of the gas, which depends on the specific heat ratio and wall temperature T_w in the expansion slot:

$$T_{tr} = \left(\frac{6\gamma}{\pi + 6\gamma}\right) T_w \tag{5.11}$$

Combining the above equations in the general expressions for the thrust and specific impulse, as given by Eqs. (5.3) and (5.4), it is finally possible to obtain the following expressions for the vacuum thrust ($p_a = 0$) and the vacuum specific impulse:

$$F_{T\_vac} = \alpha p_0 A_e \frac{\pi + 2}{2\pi} \sqrt{\frac{T_w}{T_0}\left(\frac{6\gamma}{\pi + 6\gamma}\right)} \tag{5.12}$$

$$I_{sp\_vac} = \frac{\pi + 2}{g_0} \sqrt{\frac{kT_w}{2\pi m_a}\left(\frac{6\gamma}{\pi + 6\gamma}\right)} \tag{5.13}$$

Eqs. (5.12) and (5.13) confirm that, even in the very different fluid dynamic conditions of a LPM microthruster, similar considerations as in a conventional thruster based on continuum flow apply: the thrust level is proportional to the plenum pressure and the expansion slot area, while the specific impulse is mainly dependent on the wall temperature and the molecular mass. A noticeable difference is however that the wall temperature also influences in a direct way the thrust level, while it has an almost negligible effect on the thrust level of a conventional VLM microresistojet, see Eqs. (5.1) and (5.2) combined with Eq. (5.3). Finally, the transmission coefficient has direct influence on the thrust but does not affect directly the specific impulse.

5.5 Conclusion and future challenges

To better put in the right context electrothermal microthrusters, and more specif-ically microresistojets, it is possible to compare their current state-of-the-art perfor-mance with other types of micropropulsion systems. The information presented in Figs. 5.14–5.16 is based on data collected from various review papers [4,41–46] and provides an overview of the available micropropulsion alternatives in terms of thrust, specific impulse, input power, and thrust-to-power ratio.

From this comparison, it is clear that microresistojets offer the same versatility of cold gas thrusters in terms of their available range of thrust levels, but at better spe-cific impulse (double to triple with respect to what is available with cold gas thrusters). This, of course, comes at the cost of the input electrical power required to heat the propellant.

While Fig. 5.14 clearly shows that microresistojets are still significantly far away in terms of specific impulse from other electric propulsion options and therefore not competitive with their propellant consumption efficiency, it can be clearly seen from Figs. 5.15 and 5.16 that microresistojets are by far the option with the highest thrust-to-power ratio, among those in which electrical power is needed to provide the required thrust energy to the propellant. This means that, although, with relatively poor propellant consumption efficiency, microresistojets are the option that allows for generating the highest thrust level with a given input power.

Concluding, the ideal application of miniaturized electrothermal thrusters is one in which versatility is an asset, both from the point of view of thrust and propellant

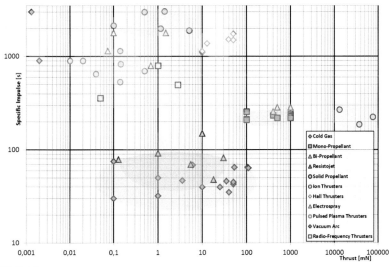

FIGURE 5.14

Current state of the art of various types of micropropulsion systems, in terms of thrust and specific impulse.

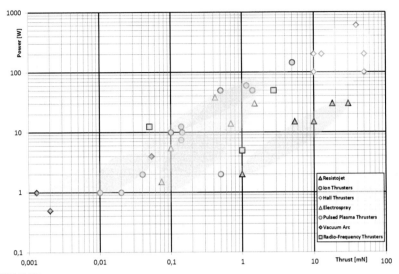

FIGURE 5.15

Current state of the art of various types of micropropulsion systems, in terms of thrust and input power.

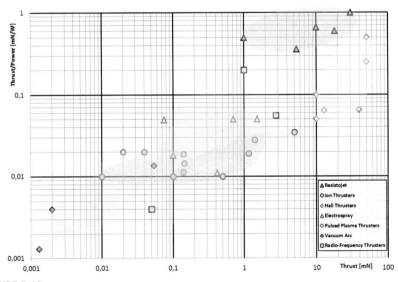

FIGURE 5.16

Current state of the art of various types of micropropulsion systems, in terms of thrust and thrust-to-power ratio.

choice. Whenever thrust levels in the order of 1−100 mN are required, in particular, microresistojets represent the only option that allows for achieving this thrust level with a power level of 10 W or lower (typical order of magnitude of the available average power in a small satellite such as a 3U CubeSat).

As it has been shown in this chapter, the state of the art of electrothermal micropropulsion is extremely dynamic, with several options and concepts currently under development in universities, research centers, and companies.

The main advantage of this type of propulsion is its versatility, in particular the fact that it can be used with virtually any propellant in any state, solid, liquid, or gaseous.

In turn, the main challenges in the design of a conventional microresistojet thruster, where propellant is vaporized and subsequently accelerated in a convergent-divergent nozzle, are the flow instabilities and pressure oscillations intrinsically present in the two-phase flow in the heating chamber, with their consequences in terms of controllability of the thruster; the thermal efficiency of the propulsion system, intrinsically limited by the small size and the materials used; the flow losses in the nozzle, which can become significant at values of throat Reynolds number lower than 1000, which are typically obtained when the thrust level is in the mN range. These drawbacks can be partially overcome by considering alternative microresistojet concepts, such as the use of a separate vaporization chamber to reduce the issues generated by the two-phase flow, or the LPM option in which the low pressure and rarefied flow regime allow for making the conventional convergent-divergent nozzle not necessary anymore.

Generally speaking, miniaturized electrothermal propulsion presents interesting advantages that make it definitely worth further research, but is currently still at an embryonal stage. More developments in the field are certainly expected in the close future, with a concrete possibility for competitive technology being developed within the next 5−10 years.

References

[1] NASA Goddard Spaceflight Center, Vela 5A, https://heasarc.gsfc.nasa.gov/docs/heasarc/missions/vela5a.html.
[2] Mickelsen, W.R., Isley, W.C., 1968. Auxiliary electric propulsion − status and prospects. In: Proceedings of the 5th Symposium on Advanced Propulsion Concepts.
[3] Aeroject Rocketdyne, 2019. Electric Propulsion Systems. Available at: www.rocket.com.
[4] Lemmer, C., 2017. Propulsion for CubeSats. Acta Astronaut. 134, 231−243.
[5] Palmer, K., Li, Z., Wu, S., 2016. In-orbit demonstration of a MEMS−based micropropulsion system for CubeSats. In: Small Satellite Conference (Logan, Utah, USA).
[6] Hejmanowski, N.J., et al., 2015. CubeSat high impulse propulsion system (CHIPS). In: Proceedings of the 62nd JANNAF Propulsion Meeting.
[7] Hejmanowski, N.J., et al., 2015. CubeSat high impulse propulsion system (CHIPS) design and performance. In: Proceedings of the 63rd JANNAF Propulsion Meeting.

[8] Hruby, V., 2012. High Isp CubeSat Propulsion, 1st Interplanetary CubeSat Workshop, Cambridge, MA, USA.

[9] Moore, G., et al., 2010. 3D printing and MEMS propulsion for the RAMPART 2U CubeSat. In: Small Satellite Conference (Logan, Utah, USA).

[10] Asakawa, J., et al., 2017. Development of the water resistojet propulsion system for deep space exploration by the CubeSat: EQUULEUS. In: Small Satellite Conference (Logan, Utah, USA).

[11] Asakawa, J., et al., 2019. AQT-D: demonstration of the water resistojet propulsion system by the ISS-deployed CubeSat. In: Proceedings of the 33rd Annual AIAA/USU Conference on Small Satellites.

[12] Koizumi, H., et al., 2019. Assessment of micropropulsion system unifying water ion thrusters and water resistojet thrusters. J. Spacecraft Rockets 56 (5), 1400−1408.

[13] Maurya, D., et al., 2005. Silicon MEMS vaporizing liquid microthruster with internal microheater. J. Micromech. Microeng. 15 (5).

[14] Kundu, P., et al., 2012. Design, fabrication and performance evaluation of a vaporizing liquid microthruster. J. Micromech. Microeng. 22 (2).

[15] Ye, X., et al., 2001. Study of a vaporizing water microthruster. Sensors Actuators A Phys. 89 (1−2).

[16] Mueller, J., et al., 1997. Design, analysis and fabrication of a vaporizing liquid microthruster. In: 33rd AIAA/ASME/SAE/ASEE Joint Propulsion Conference and Exhibit.

[17] Mueller, J., et al., 1998. Proof-of-concept demonstration of a vaporizing liquid microthruster. In: 34th AIAA/ASME/SAE/ASEE Joint Propulsion Conference and Exhibit.

[18] Mukerjee, E., et al., 2000. Vaporizing liquid microthruster. Sensor. Actuator. A 83 (1).

[19] Silva, M.A.C., et al., 2017. Vaporizing liquid microthrusters with integrated heaters and temperature measurement. Sensor. Actuator. A Phys. 265.

[20] Silva, M.A.C., et al., 2018. A comprehensive model for control of vaporizing liquid microthrusters. IEEE Trans. Control Syst. Technol. 99, 1−8.

[21] Pallichadath, V., et al., 2019. In-orbit micro-propulsion demonstrator for pico-satellite applications. Acta Astronaut. 165, 414−423.

[22] Karthikeyan, K., et al., 2012. Low temperature Co-fired ceramic vaporizing liquid microthruster for microspacecraft applications. Appl. Energy 97.

[23] Cheah, K., et al., 2015. Fabrication and performance evaluation of a high temperature Co-fired ceramic vaporizing liquid microthruster. J. Micromech. Microeng. 25 (1).

[24] Chen, C.C., et al., 2010. Simulation and experiment research on vaporizing liquid micro-thruster. Sensor. Actuator. A Phys. 157 (1).

[25] Cen, J., et al., 2010. Performance evaluation and flow visualization of a MEMS based vaporizing liquid micro-thruster. Acta Astronaut. 67, 468−482.

[26] De Giorgi, M.G., et al., 2019. A novel Quasi-one-dimensional model for performance estimation of a vaporizing liquid microthruster. Aero. Sci. Technol. 84, 1020−1034.

[27] De Giorgi, M.G., et al., 2019. Preliminary evaluation of a MEMS-based water propellant vaporizing liquid microthruster for small satellites. In: XXV International Congress, Italian Association of Aeronautics and Astronautics.

[28] Ketsdever, A., et al., 1998. A free molecule micro-resistojet: an interesting alternative to nozzle expansion. In: 34th AIAA/ASME/SAE/ASEE Joint Propulsion Conference and Exhibit.

[29] Ketsdever, A., et al., 2001. Gas-surface interaction model influence on predicted performance of microelectromechanical system resistojet. J. Thermophys. Heat Tran. 15 (3), 302−307.

[30] Ketsdever, A., et al., 2005. Performance testing of a microfabricated propulsion system for nanosatellite applications. J. Micromech. Microeng. 15 (2), 2254−2263.

[31] Ahmed, Z., et al., 2005. Numerical analysis of free molecule micro-resistojet performance. J. Propul. Power 22 (4), 749−756.

[32] Lee, R.H., et al., 2007. Performance characterization of free molecule micro-resistojet utilizing water propellant. In: 43rd AIAA/ASME/SAE/ASEE Joint Propulsion Conference and Exhibit.

[33] Guerrieri, D.C., et al., 2017. Fabrication and characterization of low-pressure micro-resistojets with integrated heater and temperature measurement. J. Micromech. Microeng. 27, 12.

[34] Guerrieri, D.C., 2016. Analysis of non-isothermal rarefied gas flow in diverging micro-channels for low pressure micro-resistojets. ASME J. Heat Transfer 138, 11.

[35] Cervone, A., et al., 2015. Conceptual design of a low-pressure micro-resistojet based on a sublimating solid propellant. Acta Astronaut. 108, 30−39.

[36] Leverone, F., et al., 2019. Cost analysis of solar thermal propulsion systems for micro-satellite applications. Acta Astronaut. 155, 90−110.

[37] Kennedy, F.G., et al., 2002. Preliminary design of a micro-scale solar thermal propulsion system. In: 38th AIAA/ASME/SAE/ASEE Joint Propulsion Conference and Exhibit.

[38] Sahara, H., et al., 2003. Single-crystal molybdenum solar thermal propulsion thruster. Trans. Jpn. Soc. Aeronaut. Space Sci. 46, 180−185.

[39] Guerrieri, D.C., et al., 2017. Selection and characterization of green propellants for micro-resistojets. ASME J. Heat Transfer 139, 10.

[40] Guerrieri, D.C., et al., 2018. An analytical model for characterizing the thrust performance of a low pressure micro-resistojet. Acta Astronaut. 152, 719−726.

[41] Levchenko, I., et al., 2018. Space micropropulsion systems for cubesats and small satellites: from proximate targets to furthermost frontiers. Appl. Phys. Rev. 5 (1).

[42] Tummala, A.R., et al., 2017. An overview of cube-satellite propulsion technologies and trends. Aerospace 4 (4), 58.

[43] Krejci, D., et al., 2018. Space propulsion technology for small spacecraft. Proc. IEEE 106 (3), 362−378.

[44] Silva, M.A.C., et al., 2018. A review of MEMS micropropulsion technologies for CubeSats and PocketQubes. Acta Astronaut. 143, 234−243.

[45] Parker, K.I., 2016. State-of-the-Art for small satellite propulsion systems. In: Proceedings of the 4th NSBE Aerospace Systems Conference.

[46] Leomanni, M., et al., 2017. Propulsion options for very low Earth orbit microsatellites. Acta Astronaut. 133, 444−454.

Further reading

[1] Sutton, G.P., Biblarz, O., 2001. Rocket Propulsion Elements, seventh ed. John Wiley & Sons.

Electrostatic microthrusters

6

Gabe Xu, PhD [1], Kristina Lemmer, PhD [2]

[1]*Associate Professor, Mechanical and Aerospace Engineering, University of Alabama in Huntsville, Huntsville, AL, United States;* [2]*Associate Professor, Department of Mechanical and Aerospace Engineering, Western Michigan University, Kalamazoo, MI, United States*

6.1 Background

Electrostatic propulsion systems are arguably the most well-known electric propulsion system today. The use of gridded ion engines in deep space science missions such as Deep Space 1 (NASA), Dawn (NASA), Hayabusa 1 and 2 (JAXA), SMART-1 (ESA), BepiColombo (ESA), and DART (NASA) has entered them into the public awareness. The three main types of electrostatic thrusters are gridded ion engines, Hall effect thrusters, and electrospray thrusters. Examples of each type are shown in Fig. 6.1. They all utilize a high electric field to accelerate ions, usually positively charged ions, to high velocities on the order of multiple km/s. However, their mass flow rates are very low, on the order of μg/s-mg/s, especially for miniature or microthrusters. The high exit velocity results in miniature thrusters with very high I_{sp}, 1000 s+, but very low thrust, <10 mN.

The first recorded discussion of the concept of ion acceleration for propulsion comes from the father of modern rocketry himself, Konstantin Tsiolkovsky. In his 1911 text, he briefly discusses the potential to "use electricity to produce a large velocity for the particles ejected from a rocket device." In the next sentence of his text, Tsiolkovsky mentions cathode rays and how electrons can obtain velocities of 30,000–100,000 km/s. At the time, only electrons were known to achieve high

FIGURE 6.1

(Left) NASA's NEXT gridded ion engine [1]; (middle) NASA HERMeS Hall effect thruster [2]; (right) Busek ionic liquid electrospray thruster [3].

Left: https://www.nasa.gov/centers/glenn/multimedia/imagegallery/if_89_ionthruster.html#.YdzRdP5BzIU.

Middle: https://www.nasa.gov/image-feature/hall-effect-rocket-with-magnetic-shielding-hermes-technology-development-unit-1/. Right: https://www.nasa.gov/sites/default/files/thumbnails/image/ieps_nasa.jpg.

Space Micropropulsion for Nanosatellites. https://doi.org/10.1016/B978-0-12-819037-1.00011-6

151

velocities, and the concept of the much heavier positively charged ion was still in debate. Tsiolkovsky's contemporary, Robbert Goddard also gave thought to the idea of using electricity for space propulsion. He first realized the need for plume neutralization in 1907. Then, in 1913 and 1917, Goddard filed two patents that formed the first technical concepts for an electrostatic accelerator. The first in 1913 described a method to produce electrically charged particles by using magnetic fields to confine electrons in a gas and thereby increase the probability of ionization collisions, which is very similar to magnetic confinement used in modern electron-bombardment gridded ion engines. In the second patent, in 1917, Goddard described three apparatus to produce a jet of electrified gas. That last design was the first documented electrostatic ion accelerator for propulsion. There are many more pioneers of electric propulsion who have contributed to science and technology. The reader is directed to papers on the history of electric propulsion [4].

The first spacecraft to use an electrostatic accelerator or gridded ion engine is the Space Electric Rocket Test (SERT-1) mission by NASA in 1964. SERT-1 carried two types of gridded ion engines, a cesium contact engine and a mercury electron-bombardment engine. The cesium engine failed to start due to an electrical short. The mercury engine was successfully operated for 31 min. The SERT-1 mercury engine was an electron-bombardment type developed at NASA Lewis Research Center, now NASA Glenn Research Center. Harold Kaufman, a physicist at NASA Lewis led the development of the engine in the 1950s and 1960s; thus, electron-bombardment gridded ion engines are at times referred to as "Kaufman or Kaufman-type ion thrusters". The development of electrostatic thrusters continued heavily in both the United States and the former Soviet Union throughout the 1960s and 1970s. The United States focused its efforts on gridded ion engines, while the Soviet Union focused development on Hall effect thrusters (HET). The first HET in space was flown on board a Soviet Meteor spacecraft in 1971. HETs were introduced to the United States in 1992 when a team of electric propulsion scientists and engineers from NASA and the US Air Force visited their counterparts in Russia to evaluate the SPT-100 HET. Today, hundreds of gridded ion engines and HETs have been flown, primarily for north–south station-keeping maneuvers on Earth-orbiting satellites. A handful have been used on deep space science and exploration missions as previously mentioned. With the advent of small satellites and the growing capability to launch them, miniature electrostatic thrusters are being studied to provide high I_{SP} in-space propulsion for a range of possible missions.

6.2 Principle of operation

Electrostatic thrusters can be broken down into three main functions: plasma generation, ion acceleration, and beam neutralization, shown schematically in Fig. 6.2 for a gridded ion engine, Hall effect thruster, and electrospray thruster. Assuming the propellant is a neutral gas, the first function of the thruster is to ionize the propellant to generate charged ions. The electrospray thruster is unique in this function as its

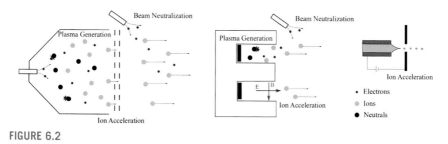

FIGURE 6.2

Main functions of three types of electrostatic thruster. From left to right: gridded ion engine, Hall effect thruster, and electrospray thruster.

Author's original.

propellant is in a liquid form, either an ionic liquid or liquid metal, that naturally contains positive and/or negative particles. Thus, there is no traditional ionization or plasma generation in electrospray thrusters. Alternatively, one can think of the liquid propellant as already being in an ionized state containing charged particles.

After ionization and plasma formation, the second function of the thruster is to accelerate the charged particles to high velocities using a large electric field. In gridded ion engines, this is a pair of aligned grids, a screen grid and an accelerator grid, called ion optics. The two grids are biased at different voltages to create a voltage change on the order of kilovolts. With the grid spacing on the order of a few millimeters, this creates a very strong electric field for ion acceleration. In the case of an HET, the ion acceleration occurs not by physical grids, but due to the existence of a radial magnetic field that traps electrons. The field reduces the electron's cross-field mobility, or ability to reach the positively biased anode. This causes the applied voltage between the anode and cathode to be distributed along the axis of the channel, generating an axial electric field for acceleration. The magnetic field typically has a maximum strength near the exit plane of the thruster and decreases to near zero at the anode. This has the effect of compressing the maximum voltage change; thus, the largest electric field occurs near the regions of high magnetic field at the exit.

The last main function of an electrostatic thruster is beam neutralization. One of the main differences between electrostatic thrusters and other types of electric propulsion is the need for beam neutralization because only one type of charged particles, positive or negative ions, can be ejected at one time through a single exit. When the thruster only emits one charge, for example positive ions, it and the associated spacecraft will quickly become strongly negatively charged. This could cause serious electrical hazards and shorts that will damage the spacecraft. To prevent charge build-up, electrostatic thrusters eject free electrons, in the case of a thrust beam made of positive ions, into the plasma plume to maintain net system charge balance. The most common electron sources are hollow cathodes or simple tungsten filaments. Researchers have also developed clever methods to circumvent the need for a separate electron source. For example, the use of two thrusters wherein one accelerates positive particles and the other accelerates negative particles. Another

interesting solution is the PEGASES thruster which uses a propellant that can produce positive and negative ions (SF_6). The accelerator grid voltage can be switched
from positive to negative to alternately extract positive and negative ions, removing
the need for a separate neutralizer. This concept is also how many ionic liquid electrospray thrusters neutralize their plume. The polarity on the extractor grid is
switched so that anions and cations are alternatively accelerated.

The theory for electrostatic thrusters has been well established from work done
to date on large-scale devices, and this work is applicable to miniature thrusters.
However, additional considerations and inefficiencies arise at a small scale, which
will be discussed.

6.2.1 Ionization and plasma generation

As electrospray thrusters use a liquid propellant with existing ions, they do not have
an ionization and plasma generation component. For gridded ion engines and HETs,
the energy efficiency of the plasma generator is critical for operation of the thruster.
There are three primary methods of gas ionization for electrostatic propulsion being
used or researched: DC electron-bombardment, radio frequency (RF) ionization, and
electromagnetic surface waves.

Fig. 6.2 showed simplified schematics of a DC electron-bombardment gridded
ion engine discharge chamber and HET discharge channel. In electron-
bombardment ionization, the discharge volume is filled with the neutral propellant
gas. Then, high-energy electrons, sufficiently greater than the first ionization energy
of the propellant species, are injected into the volume. Some of the electrons will
collide with gas atoms with sufficient energy to remove a valence electron, producing a positively charged ions and two electrons. The ion production rate is given by
Eq. (6.1).

$$\frac{dn_i}{dt} = n_e n_0 \langle \sigma_i v_e \rangle \tag{6.1}$$

Here, n_i, n_e, and n_0 are the number densities of ions, electrons, and neutral atoms,
respectively. σ_i is the ionization cross-section and v_e is the electron velocity. The
term in $< \dots >$ is the ionization rate and the brackets mean that the cross-section
is averaged over the electron velocity distribution function. For a xenon gas with
an average electron energy of 10 eV, the ionization rate is approximately
3.9×10^{-14} m^3/s [5]. While this appears like a small number, given the electron
and neutral densities in thrusters are on the order of 10^{16} and 10^{18} m^3/s, respectively,
the ion production rate is $\sim 10^{20}$ m^3/s. However, this is not the complete story for
plasma generation. The ion production is countered by ion loss due to recombina-
tion, ion extraction for thrust, and wall neutralization. The latter term dominates
as the volume-to-area ratio of the discharge chamber or channel decreases, as the
case in miniature thrusters. To improve the probability of energetic electrons
colliding with neutral atoms, electron-bombardment gridded ion engines and
HETs use magnetic fields to confine electrons. Gridded ion engines traditionally

use a set of cusp magnetic fields around the boundaries of the discharge chamber at sufficient magnetic field strength to reflect electrons. HETs trap electrons on radial magnetic field lines.

Radiofrequency ionization is a technique used in RF gridded ion engines. Instead of a discharge cathode that provides high-energy electrons for impact ionization, a set of RF coils are wrapped outside the discharge chamber. The coils are driven at a low-frequency RF signal, typically at 13.56 MHz. With the coils outside the chamber, an inductively coupled plasma (ICP) is produced. External coils and ICPs are preferred as the configuration prevents plasma sputtering and damage to the RF coils, thus extending the lifetime of the thruster. In ICPs, the RF current cannot directly interact with the plasma, resulting in the "inductive" name. Instead, the RF current produces a time-varying magnetic field outside the coil. In turn, the time-varying magnetic field generates an azimuthal electric field in the discharge chamber according to Maxwell's equations. This electric field accelerates electrons to initiate collisional ionization and an electron avalanche that creates the plasma. Once the plasma is created, the time-varying magnetic field generates eddy currents in the plasma in the opposite direction to the current in the RF coil. The eddy currents generate opposeing magnetic fields to shield some of the plasma from the applied field. Therefore, the RF power can only penetrate a certain distance into the plasma, called the skin depth. The skin depth is a function of pressure and RF frequency. For RF gridded ion engines, it is desirable for the skin depth to be on the order of the radius of the discharge chamber for optimal RF energy deposition.

An often forgotten method of ionization tested for electrostatic thrusters is the contact ionization used on the failed cesium contact engine aboard SERT-1. In contact ionization, a low work function fluid, such as liquid cesium, is passed through a heated porous tungsten bed. The higher work function of tungsten compared to the cesium causes electron transfer from the cesium to the tungsten producing cesium ions. The vaporized cesium ions then are electrostatically accelerated to produce thrust. Cesium contact engines were tested multiple times from 1968 to 1974 on the ATS-4 through ATS-6 satellites.

6.2.2 Ion acceleration

For Gridded ion engines, the most common type of ion optics are a two-grid system, though three-grid systems also exist. Electrospray thrusters typically only uses a single extraction grid electrode, and HETs use the magnetized electrons to generate the electric field. The ion exit velocity can be calculated from conservation of energy between the ion's kinetic energy and the electrical energy as

$$\frac{1}{2}m_i v_i^2 = qV_b \tag{6.2}$$

$$v_i = \sqrt{\frac{2qV_b}{m_i}} \tag{7.3}$$

where m_i and v_i are the ion mass and exit velocity, q is the ion charge, and V_b is the beam voltage through which the ion was accelerated. Thus, the calculation of electrostatic performance is relatively straightforward provided one knows the charge state and voltage. As Eq. (6.3) shows, there is an inverse relationship between the particle mass and velocity. For the same accelerating beam voltage, V, low-mass particles such as hydrogen or helium would reach a much higher exit velocity than high-mass particles such as xenon. Since specific impulse is a direct function of exit velocity, given by $I_{SP} = v_e/g_0$, where $g_0 = 9.81$ m/s^2, the lighter species are desirable for high I_{SP}. On the other hand, the thrust produced scales with particle mass via

$$T = I_b \sqrt{\frac{2m_i V_b}{q}} \tag{6.4}$$

where I_b is the ion beam current. Eq. (6.4) is a result of the ion mass flow rate being a function of the beam current, $\dot{m} = I_b m_i \big/ q$. When substituted into the rocket equation along with Eq. (6.3), one gets the form of Eq. (6.4). Comparing Eqs. (6.3) and (6.4), it is clear that particle mass has opposite effects on I_{SP} (via exit velocity) and thrust.

6.2.3 Beam neutralization

To date, electrostatic thrusters primarily accelerate positively charged ions; therefore, the beam neutralizer must provide electrons. To prevent charge build-up, the neutralizer current must equal the ion beam current. The SERT-1 gridded ion engine used the simplest of electron sources, a heated tantalum wire, akin to that used in incandescent light bulbs. Filament neutralizers work on the simple principle of thermionic emission. Thermionic emission is the process by which a material will emit charge carriers when sufficiently heated. This occurs because thermal energy can overcome the work function of the material. The work function is the minimum energy needed to remove an electron from the surface of a solid. While all solids have a work function, most have melting points below the work function temperature. The emitted current density, J, from a material can be calculated using the Richardson equation, also known as the Richardson–Dushman equation.

$$J = AT^2 e^{\frac{-e\phi}{kT}} \tag{6.5}$$

In the equation, A is a theoretical constant (120 A/cm^2-K^2), T is the temperature in K, e is the elementary charge, ϕ is the work function in Joules, and k is Boltzmann's constant (1.38×10^{-23} J/K). The constant A has been experimentally found to vary with temperature, thus a modified Richardson equation can be used with a material specific constant, D, and a temperature modified work function $\phi = \phi_0 + \alpha T$, where ϕ_0 is the classical work function and α is an experimentally determine constant.

$$J = DT^2 e^{\frac{-e\phi_0}{kT}} \tag{6.6}$$

From these equations (the constants are summarized for common thermionic materials used in electric propulsion by Ref. [6]), it is possible to determine the length of filament needed for a given filament material to achieve the electron current needed to match the ion beam current.

While filament neutralizers are simple, they have difficulty providing high electron currents over 1 A without impractical filament lengths. For high currents, the most common neutralizer is a hollow cathode neutralizer. The hollow cathode consists of a refractory metal cylinder with a cylindrical thermionic material inside. A hollow cathode extracts the electron current from a plasma inside the cathode and is capable of very high currents. The cathode plasma is created by the ionization of the propellant gas flowing through the tube. This cathode gas flow is a fraction of the main thruster gas flow, on the order of 1%−3%. While this flow is necessary for the operation of the hollow cathode, it is not accelerated to produce thrust and should be considered a performance loss. Ionization electrons are created by electron emission from the thermionic material following the Richardson−Dushman equation. To achieve the temperatures necessary for thermionic emission, either a resistive heater coil is wrapped outside the cathode tube, or an arc plasma is established inside the cylinder. Both have benefits and drawbacks. A resistively heated cathode requires a dedicated multiampere heater power supply and takes 10s of minutes to go through a heating schedule before the thermionic material can emit electrons. High currents in the heater coil slowly wear down the coil and can lead to cathode failure. Arc heated cathodes, also called arc start or instant start cathodes, can initiate electron emission in less than a minute, and in seconds in some cases. However, the high current arc that heats the thermionic material can sputter and damage the thermionic material leading to failure.

As the thermionic electrons are emitted and ionize the gas, a sheath forms on the surface of the thermionic material. The sheath is a voltage or potential boundary layer that accelerates ions toward the surface and electrons away from the surface. This way, the emitted electrons are accelerated after emission with sufficient energy to ionize the gas. Plasma ions in turn are accelerated to the surface and provide collisional heating of the thermionic material. At sufficiently high currents and cathode plasma densities, the ion bombardment is large enough to maintain the temperature of the thermionic material and the hollow cathode becomes self-heating, at which point the external heater can be turned off. The electrons from the cathode plasma are extracted through an orifice at the end of the cathode cylinder. The size of the orifice varies and typically increases with the extracted current. Electrically, the orifice, cathode body, and thermionic material are all at the same potential, which is far below the anode potential in the thruster. This allows electron current extraction, but can also attract ions from the thruster plume that will damage the orifice and cathode material. Therefore, a keeper or extraction electrode is typically placed outside the orifice. This often takes the form of another orifice plate but with a large hole. This keeper electrode is electrically biased slightly positive. The keeper helps initiate the cathode discharge by extracting a small electron current and protects the orifice from direct ion bombardment. The keeper will experience ion bombardment, and the keeper hole will grow over time, eventually leading to failure.

Hollow cathodes have been developed and tested for electron currents in the hundreds of Amperes range. To get to such high currents requires larger cathodes with larger inserts. This is driven by the fact that electrons are only emitted from the surface of the thermionic material. Thus, if the thermoionic material can emit at a current of $J = 20$ A/cm^2, 10 A of electron current requires only 0.5 cm^2 of emitter surface area. However, to emit 500 A of electron current will require 25 cm^2 of emitter surface area. If we assume the same diameter of the thermionic cylinder, this is a 50× increase in thermionic emitter length. The other consideration for high current hollow cathodes is space charge limiting the thermionic emission. The emitted electron density that can be accepted by the plasma is given by

$$J_e = \frac{1}{4}n_e e \sqrt{\frac{kT_e}{m_e}}$$

Thus, for the 20 A/cm^2 case, to prevent space charge limited emissions, one would need an electron density in the plasma of 1×10^{13} cm^3 for an electron temperature of 1 eV.

6.3 Selection of propellant

Propellants for electric propulsion (EP) systems include all three phases of matter: solid, liquid, and gas. For systems that require ionized plasma, such as gridded ion engines and HETs, a gas phase is required for the propellant. However, some solids and liquids with relatively low vaporization temperatures have been used. The earliest propellants used for in-space thrusters were cesium and mercury on board the SERT-1 mission. Today, flight gridded ion engines and HETs typically use xenon propellant, with the exception of the SpaceX Starlink satellite HETs that use krypton. Electrospray thrusters use either a liquid metal or ionic liquid as a propellant, but, there is no need for ionization and plasma generation in electrospray thrusters. Modern solid condensable propellants include elements such as iodine, bismuth, and magnesium.

6.3.1 Gaseous

Xenon is the most commonly used propellant in gridded ion engines and HETs. The choice is based on three factors of (1) ionization energy, (2) atomic mass, and (3) safety. The largest energy sink in a gridded ion engine or HET is the energy or power required to produce the discharge plasma. Magnetic fields in these thrusters help reduce the loss of energetic electrons before they can have an ionizing collision, reducing the waste power, but a major driver of the ionization cost is the innate ionization energy of the propellant. Thus, the lower the first ionization energy of a propellant, the lower the power required to produce the plasma and ions for acceleration. The second factor, atomic mass, affects the thrust produced as a higher atomic mass will result in higher thrust, as previously discussed. Thrust from an EP

Table 6.1 The first ionization energy, atomic mass, and safety considerations for some electric propulsion propellants.

Gas	1st ionization energy (eV)	Atomic mass (g/mol)	Safety
Helium	24.59	4.003	Inert
Argon	15.76	39.948	Inert
Krypton	13.99	83.798	Inert
Xenon	12.13	131.293	Inert
Radon	10.75	222.018	Radioactive
Mercury	10.40	200.590	Toxic to humans
Cesium	3.89	132.905	Highly reactive with air and water

system can be simply calculated from Eq. (6.4). While a primary advantage of EP is the high I_{SP}, certain orbit maneuvers require a minimum level of thrust force. For orbit transfers or interplanetary flight, the higher the thrust, the faster the trip. The last factor to consider in propellant selection is safety. Xenon eventually came to replace mercury due to the human health hazards of handling mercury. Table 6.1 gives some of the data for the naturally occurring noble gases along with mercury and cesium.

The table makes clear why xenon is generally the preferred gaseous propellant. It has the lowest ionization energy and highest atomic mass of inert noble gases. It also shows why mercury and cesium were tested in the beginning, despite safety and handling hazards. Argon and krypton have also been tested with gridded ion engines and HETs. They produce high I_{SP}, lower thrust, and require more power per ion to generate the plasma. The Starlink constellation satellites by SpaceX use krypton propelled miniature HETs becuase krypton is significanlty less expensive than xenon, and the increase in I_{sp} from the lower atomic mass is beneifcial to maintain a LEO constellation orbit.

6.3.2 Liquid

The first gridded ion engine flown aboard SERT-1 used liquid mercury as its propellant. In that application, the mercury was first vaporized and then ionized in the plasma generator. Since then, gridded ion engines and HETs have largely switched to noble gas propellants. Liquid propellants are used today in electrospray thrusters, though in very different ways. The design of electrospray thrusters requires a liquid propellant with ionic components, necessitating the use of either ionic liquids or liquid metals. Ionic liquid thrusters are sometimes called colloid or just electrospray thrusters, and liquid metal thrusters are often called field emission electric propulsion (FEEP).

An ionic liquid is a molten, or liquid salt. While all salts can reach a molten state at high temperatures, the types used in propulsion are room temperature ionic liquids. Unlike saltwater, where the salt is dissolved in a solution, in ionic liquids, the salt molecule is the liquid itself. There are many different man-made ionic liquids that have been tested for electric propulsion [7−9]. The first used in space was 1-ethyl-3-methylimidazolium bis(trifluoromethylsulfonyl)imide, or [Emim][Im], on the Laser Interferometer Space Antenna (LISA) Pathfinder mission. Ionic liquid electrospray does not require an external electron neutralizer as the polarity of the extraction grid can be reversed to alternatively extract the positive and negative salt ion molecules.

FEEP thrusters use liquid metal as a propellant. These metals are typically solid at room temperature and become liquid upon heating. Indium and cesium have been used for FEEP thrusters due to their low melting point. Unlike ionic liquid electrospray thrusters, FEEPs only extract the positive metal ions, thus still requiring a plume neutralizer.

6.3.3 Solid

The first solid propellant used in EP is likely lithium in the Lorentz force accelerator (LFA). The LFA is a type of electromagnetic propulsion device that uses the Lorentz, JxB, force to accelerate the plasma to produce thrust. As a gaseous plasma was still needed, the lithium was heated and vaporized before injection into the thruster. Since then, other condensable propellants have been tested in the lab, and the first was flown in 2021 onboard the Beihangkongshi-1 satellite with the NPT30-I2 iodine miniature gridded ion engine by ThrustMe. Condensable propellants are gases that are solid at room temperature. Elements tested include iodine [10−13], bismuth [11,14,15], zinc [11,16], and magnesium [11,16]. The main benefit of condensable propellants is their high solid density, which reduces the volume and mass required for propellant tanks over compressed xenon. For example, solid iodine has a density of 4.933 g/cm^3. Xenon, in its highly cooled and compressed cryogenic liquid state has a density of only 2.942 g/cm^3. Flight systems do not store xenon in the cryogenic state due to lack of adequate cooling in space. Common tankage fractions, which is the fraction of xenon mass to propellant tank mass, can vary from as low as 2.5% −14% [17−19]. Additionally, a condensable propellant puts fewer requirements on the tankage as it can be stored at ambient pressure and temperature with only a heater system required to vaporize the solid propellant.

Condensable propellants present two main challenges: surface coating and flow control. To keep the propellant particles in the gas state requires maintaining the gas above the vaporization temperature. In the propellant tank and feed lines, this is done using resistive heaters. After the ions are accelerated out of the thruster, they will quickly cool and return to their solid state. This presents a problem for both ground test facilities and in-space operations. In ground test vacuum chambers, the propellant will coat the inside of the chamber and potentially the vacuum pumps. If the propellant is unreactive, then the chamber can be cleaned after testing. For in-space operation, the gaseous ions can condense on spacecraft surfaces, coating critical components such as solar panels, cameras, and radiators. This can be minimized by proper placement of the thruster and other satellite components such that the exhaust plume has minimal interactions with spacecraft surfaces. The surface coating issue is also a concern to the thruster system itself. When the thruster is turned off, there will be residual gaseous propellant in the lines and thruster body. Those particles will condense on the nearest surface. For the most part, this is not a concern for the tank or lines; however, it is a concern for the discharge chamber or channel, valves, and the cathode. A coating of the condensable solid in valve seats can prevent valve closing. A coating of conducting metal such as zinc on the surface of a dielectric channel HET can alter thruster behavior. Potentially the most damaging of all is coating the thermionic surface in hollow cathodes. Hollow cathodes require a clean emission material to produce the initial electrons to start the thruster. If the condensable propellant coats or chemically interacts with a sufficient portion of the emitter surface, the cathode may be rendered inoperable. For this reason, many of the experiments with condensable propellant use a separate xenon feed for the cathode and only run the condensable propellant through the main thruster. The ThrustMe iodine gridded ion engine used a filament cathode instead of a thermionic hollow cathode.

The second challenge of condensable propellants is accurate metering of the mass flow rate. The early efforts with these propellants used thermally controlled reservoir tanks to control the flow rate. This has the advantage of being simple and relatively robust but incurs higher variation and uncertainty in flow rate. For flight use, accurate metering is important for fine orbit maneuvers. Electronic mass flow controllers for iodine have been developed and tested [13,20].

6.4 Current state of the art

The classification of "miniature" or "micro" thruster is a loose one. There is not a universally accepted definition, but here we will define it as a thruster that can fit in a few "U's" of volume and uses less than ~ 200 W of power. A "U" is a cube 10 cm on each side. The terminology originates from the CubeSat size standards. As the growth of the small satellite field has only ramped up in the last 10−20 years, the need for on board propulsion is still catching up. However, that has not stopped research groups from around the world from developing and testing miniature EP systems.

6.4.1 Systems with flight heritage

Possibly the first small electrostatic thruster flown was the RIT-10 by The Ariane Group in 1992 on board the EURECA spacecraft. The RIT-10 is an RF gridded ion engine with a nominally 10 cm diameter grid that produces 5 mN with 1900 s at 145 W power. The RIT-10 was later flown again on the ARTEMIS mission in 2001.

The first miniature HET flown was the BHT-200 by Busek in 2006 on board the TACSAT-2. The thruster was subsequently flown on FalconSat-5 (2010) and FalconSat-6 (2018). It is currently part of the NASA iSAT mission to demonstrate the use of iodine propellant in HETs on orbit. The thruster produces 13 mN thrust at 200 W and is operable from 100 to 300 W.

The SPT-50 is a 200-W class HET developed by the company OKB Fakel in Russia. It was first flown on the Canopus-V mission in 2012. The SPT-50 can operate from 100 to 700 W of power and produce 7–35 mN of thrust with 800–1700 s I_{sp} [21]. The version flown produced 14 mN and 850 s at 317 W of power [22]

The Ion thruster and Cold-gas thruster Unified Propulsion System (I-COUPS) was a combined gridded ion engine and cold gas thruster developed by the University of Tokyo and flow on board the PROCYON mission in 2015. The gridded ion engine portion of I-COUPS produced 300 µN and 1080 s at 33 W of system power [23]. The cold gas thruster produced 22 mN and 24.5 s at 7 W of system power. The I-COUPS gridded ion engine is unique in that the neutralizer is not a thermionic hollow cathode, but instead is a separate gridded ion engine that extracts electrons from the plasma (Fig. 6.3).

The Ion Electrospray Propulsion System (iEPS) is an electrospray thruster using ionic liquids that were flown on the IMPACT or AeroCube8 mission in 2015. The iEPS was originally developed at MIT and later commercialized and renamed the Tiled Ionic Liquid Electrospray (TILE) by the start-up company Accion Systems. The version flow on the IMPACT mission consisted of a propulsion module with eight thruster heads, and each thruster head had 480 micro fabricated emitters [24]. The module produced 50 µN with 1250 s at 1.5 W of power. The TILE thrusters can be scaled up to higher thrust levels by adding more modules. For example, the TILE 5000 shown below is designed to produce 1.5 mN with 1500 s at 30 W of power (Fig. 6.4).

FIGURE 6.3

I-COUPS thruster head with the four-hole neutralizer above [23].

J. Rovey, C.Lyne, A. Mundahl, N. Rasmont, M. Glascock, M. Wainwright, S. Berg, Review of multimode space propulsion, Prog. Aero. Sci. 118 (2020) 100627.

FIGURE 6.4

A single iEPS module consisting of eight thruster heads [24].

M. Jenkins, D. Krejci, P. Lozano, CubeSat constellation management using ionic Liquid Electrospray Propulsion,
Acta Astronaut. 151 (2018) 243–252.

The IFM Nano Thruster is the first liquid metal electrospray, also known as FEEP, to be flown in space. The thruster is developed by FOTEC/ENPULSION. It was first demonstrated in orbit in 2018 on a CubeSat. The thruster can produce from 0.25 mN with 5000 s at 30 W up to 0.42 mN with 3200 s at 40 W [25]. Larger units can be built for greater thruster levels [26] (Fig. 6.5).

The Busek Colloid Micro-Newton Thruster was launched on the LISA Pathfinder mission in 2015. There are four emitter heads on each thruster unit, with nine needle emitters in each head. The thrust per emitter is 0.1−3 μN and between 500 and 1500 s [3].

The last flight miniature electrostatic thruster is the Starlink HET by SpaceX. At the time of writing, there is not much information about the specifications and

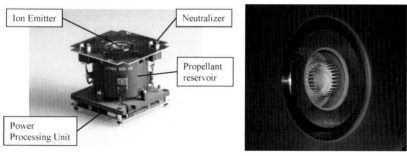

FIGURE 6.5

The IFM Nano Thruster [26] and view of the thruster head during operation [25].

D. Krejci, V. Hugonnaud, T. Schönherr, B. Little, A. Reissner, W. Neustadt, B. Seifert, Q. Koch, E. Borràs, and J.
del Amo, Full performance mapping of the IFM nano thruster, including direct thrust measurements, JoSS 8 (2)
(2019) 881–893.

performance of these thrusters. It is an annular HET operating on a krypton propellant developed by SpaceX. With over 900 Starlink satellites launched by the end of 2020 and 1440 total satellites planned for the constellation, this would be the largest use of electric propulsion in history, miniature, or otherwise.

6.4.2 Systems under development

Although only a handful of miniature electrostatic thrusters have flown, there are many thrusters that have been or are being developed by government, industry, and academia. This section will attempt to list the wide range of known miniature electrostatic thrusters that have yet to fly. The majority of these thrusters are in the laboratory stage and do not have power processing units, propellant tanks, or valve systems suitable for spacecraft integration. To limit the range of research currently in this area, we will only consider thrusters that have an assembled prototype in the lab. Research on new components of miniature electrostatic thrusters, but not incorporated in a whole device, will not be included. To organize the list, the systems are divided into gridded ion engine type, HET type, electrospray type, and others.

6.4.2.1 Miniature gridded ion engines

Following the success of the RIT-10, The Ariane Group developed a smaller version called the RIT-μX in 2007 [27]. The RIT-μX is a scaled-down version of the RIT-10 and can produce thrust levels from 50 to 500 μN with 300–3000 s at <50 W of power [28]. The thruster fits in less than 1U and weighs 440 g.

The Miniature Xenon Ion (MiXI) thruster is a 3-cm diameter gridded ion engine initially developed by the California Institute of Technology and NASA Jet Propulsion Laboratory (JPL) in 2001 [29]. The thruster development was later moved to the University of California in Los Angeles in partnership with NASA JPL [30,31]. The MiXI is a permanent magnet electron bombardment gridded ion engine with either a 3-ring or 5-ring cusp field configuration. The thruster can produce thrust in the range of 0.1–1.5 mN, 1700–3200 s I_{SP} at 14–43 W of power (Fig. 6.6) [31].

The BIT-1 and BIT-3 are small RF gridded ion engines developed by Busek Inc. The BIT-1 produces 185 μN and 1600 s at 28 W of power. The larger BIT-3 has a 2.5-cm diameter grid and produces 1.15 mN and 2100 s at 75 W of power. The BIT-3 has been developed into a complete subsystem with iodine propellant for integration into 6U CubeSats (Fig. 6.7) [32].

The Neptune gridded ion engine is an RF-powered miniature thruster developed by the company ThrustMe in France in partnership with the Laboratory of Plasma Physics at Ecole Polytechnique and SATT Paris-Saclay, a technology incubator. The thruster is unique in that the ion extraction grids are powered with RF voltage instead of DC. The oscillating grid voltage allows both ion and electron extraction from the same grid, allowing beam neutralization without a dedicated neutralizer cathode. The thruster has been tested with xenon and iodine propellant use is underway. The expected performance is 0.2–0.7 mN and >1000 s at 30–60 W of power [33]. The thruster body has been integrated into a 1U frame.

FIGURE 6.6

The MiXI thruster with a 3-cm diameter grid [31].

E. Dale, B. Jorns, A. Gallimore, Future directions for electric propulsion research, Aerospace 7 (120) (2020).

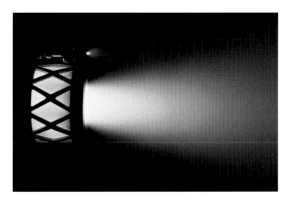

FIGURE 6.7

The integrated flight units for 6U CubeSats [32].

https://commons.wikimedia.org/wiki/File:Busek%27s_BET-3_Ion_Thruster.jpg.

A related effort to scale down the RIT-10 was done at Giessen University that resulted in the μNRIT-2.5 with a 2.5-cm diameter ion grid [34]. A number of predecessor units with different-sized grids and holes were tested as well. Like the RIT-10, the μNRIT-2.5 is an RF power gridded ion engine. It can produce 0.575 mN and 2860 s at 34.4 W of total power [34].

The μ1 gridded ion engine was developed by The University of Tokyo and JAXA. The thruster uses an RF ionization scheme at 4.2 GHz for electron cyclotron resonance (ECR). The grids have a diameter of 16 mm. Typical thruster performance was 297 μN and 1100 s at 15.1 W of total power [35]. The μ1 was tested with a conventional cathode neutralizer as well as a second μ1 with reversed polarity to extract electrons.

Pennsylvania State University developed the miniature radio-frequency ion thruster (MRIT) [36] and the miniature microwave-frequency ion thruster (MMIT) [37]. The MRIT has a discharge chamber diameter of 1 cm and operates at 1−1.5 MHz. It produced 59 μN and 5480 s using argon propellant at 15 W of RF input power. The MMIT has a chamber diameter of 2 cm and operates in ECR modes at 4.2 GHz. It produced 250 μN and 5500 s with xenon propellant using 1 W of RF power and 6 W of grid power.

The split-ring resonator ion thruster (SSRIT) was developed at The University of Alabama in Huntsville [38]. It is an RF/microwave ionization thruster operating at 890 MHz using a split-ring resonator to generate a small surface plasma. On argon propellant, the SSRIT produced a maximum of 3.46 μN and 3800 s using 20.5 W of power (Fig. 6.8).

The Miniaturized Differential Gridded Ion Thruster (MiDGIT) is one of the more unique designs that uses two sets of extraction grids on opposite sides of a discharge chamber [39]. It was developed by The University of Southampton and QinetiQ in the United Kingdom. It is an RF ionization thruster operating at 5.25 MHz with 13−18 W of forward RF power. It produced a thrust of 250−480 μN and 600−1100 s. Differential control of the two ion beams can allow fine tuning of the net thrust output.

The Lanzhou Institute of Physics in China developed the Lanzhou Radio-frequency Induced Thruster (LRIT-40) and the Lanzhou Electron Cyclotron Resonance (LECR-50) thruster [40]. The LRIT-40 is a 50-W class gridded ion engine with a 40-mm diameter grid operating at MHz frequencies. It produces 1−2 mN of thrust and 2600 s of specific impulse. The LECR-50 is a 150-W class gridded ion engine operating in the microwave range for ECR ionization. It has a 50-mm diameter grid and produces 2.3 mN and 4010 s.

FIGURE 6.8

The split-ring resonator and the surface plasma, the SSRIT, and the thruster in operation [38].

Author's original.

The final gridded ion engine type device is the Inertial Electrostatic Confinement (IEC) thruster. IEC was originally a fusion reactor concept where a pair of nested metal wire spheres could accelerate ions toward the center of the sphere to high velocities for fusion. It is commonly called a fusor. By altering the geometry of the device, one could direct the accelerated ion beam in one direction to produce thrust. Various groups have experimented with these devices [41,42], though none have reached the stage of detailed thrust performance measurements. Winter calculated a thrust of 0.75–1 mN at powers up to 60 W [42].

6.4.2.2 Miniature Hall effect thrusters

The HT-100 Hall effect thruster was developed by Alta SpA, which is now part of Sitael SpA, in Italy. The thruster is designed for <200 W of power. The performance can range from 2 to 11 mN and 300–1000 s [43].

The Halo Hall effect thruster is developed by the commercial company ExoTerra located in the United States. It can use a range of gaseous propellants and iodine. The thruster has been tested for 4–33 mN and 700–1500 s at 74–450 W of power [44].

While most HETs are annular in design, there are a number of miniature HETs that use a cylindrical design, removing the inner wall. This reduces the surface area of the walls, and thus lowers ion loss to the walls. However, without an inner wall and the inner magnetic pole, the magnetic field of cylindrical HETs are necessarily different. Such designs have been studied by the Princeton Plasma Physics Laboratory (PPPL) [45,46], the University of Toronto [47], Osaka University [48], and others [49–51]. They are generally low powered at <200 W. The PPPL thruster produced 3–6 mN and 1100–1650 s at power levels between 90 and 185 W (Fig. 6.9) [51].

The ExoMG Hall thruster is produced by the commercial company Exotrail located in France. The thruster produces 1.5 mN and 800 s at 50 W of power (Fig. 6.10) [53].

 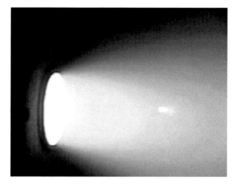

FIGURE 6.9

(Left) PPPL's cylindrical HET [52] and (right) drawing of cylindrical HET design [47].

Left and Right: A. Diallo, S. Keller, Y. Shi, Y. Raitses, S. Mazouffre, Time-resolved ion velocity distribution in a cylindrical Hall thruster: Heterodyne-based experiment and modeling, Rev. Sci. Instrum. 86 033506 (2015).

FIGURE 6.10

The ExoMG firing [54] and in a CubeSat sized unit [55].

E. Dale, B. Jorns, A. Gallimore, Future directions for electric propulsion research, Aerospace 7 120 (2020).

Stanford University developed a micro HET with an annular discharge channel width of 0.5 mm. The thruster uses samarium cobalt permanent magnets to generate the radial magnetic fields in the channel. The small channel and use of rare earth magnets produced very high magnetic fields of 0.7−1 T inside the channel. The thruster produced 0.5−1.6 mN and 300−850 s at 10−40 W of power [56].

The Divergent Cusp-field Hall Thruster (DCHT) was developed at the Massachusetts Institute of Technology [57,58]. Like the cylindrical HET, the DCHT also lacks an inner wall and center magnetic pole. The DCHT uses multiple cusp fields generated by permanent magnets to confine electrons, similar to an electron-bombardment DC gridded ion engine. Maximum efficiency was achieved at 550 V on the anode that consumed 242 W of power and produced 13.4 mN and 1650 s [58].

The Quad Confinement Thruster (QCT), developed at the University of Surrey [59,60], is a design that moves away from the traditional cylindrical or annular HET. Four sets of cusp magnetic fields are generated in a square discharge channel. An axial electric field is generated between the external cathode and the rear anode. The ExB drift drives electrons toward the center of the channel creating a high-density discharge. At 200 W anode power, the QCT produces 2−3 mN and 200−400 s with krypton propellant (Fig. 6.11) [59].

The magnetically shield miniature Hall thruster (MaSMi) is a small Hall thruster developed by UCLA and JPL and fabricated by Apollo Fusion located in California [61−63]. The thruster is part of the Ascendant Sub-kW Transcelestial Electric

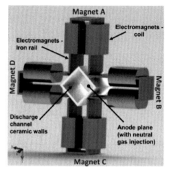 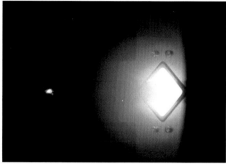

FIGURE 6.11

Quad Confinement Thruster drawings [60] and operation [59].

Left and Right: A. Fabris, C. Young, A. Knoll, E. Azevedo, M. Cappelli, Evidence of a free-space ion acceleration layer in the plume of a quad confinement plasma source, J. Appl. Phys. 131 013302 (2022).

Propulsion System (ASTRAEUS) program that seeks to develop an optimized miniature electric propulsion system for interplanetary small satellite exploration. The ASTRAEUS system includes a power processing unit, xenon flow controller, and off-the-shelf gimbal. The MaSMi is designed to operate from 150 to 1000 W of thruster power. The thruster has a performance range of 9—70 mN, 900—1600 s, and 20%—55% total efficiency at a nominal 300-V discharge voltage [62]. The magnetic shielding design refers to a particular topography of the applied magnetic field in Hall thrusters that prevents the magnetic field lines from intersecting the discharge channel walls as shown notionally in Fig. 6.12. Under the assumption that the magnetic field lines magnetize and trap electrons and are thus equipotential lines of force, this topography places low energy electrons closer to the wall and reduces the chances of high energy particles striking and eroding the discharge channel wall, which eventually would lead to the end of life of the thruster. Apollo Fusion also produces the Apollo Constellation Engine HET, but its power rating of 400 W places it outside the scope of this text.

The Aurora HET is produced by Orbion Space Technology located in Michigan. The Aurora is rated for 150—300 W power and produces 5.7—19 mN of thrust and 950—1370 s of I_{SP} [64,65]. It is also a magnetically shielded design.

A novel configuration of the standard HET is the so-named Wall-Less Hall thruster developed at the French Centre for Scientific Research (Centre national de la recherché scientifique). The wall-less configuration places the anode and the peak magnetic field at the exit of the discharge channel. This design shifts the hot ionization plasma region away from the channel walls, thereby reducing wall erosion. The wall-less design is incorporated into the 200 W ICARE small customizable thruster (ISCT-200). Performance data for the wall-less ISCT-200 are not available at the time of writing. Experiments with a larger 1.5 kW wall-less thruster showed comparable thrust and Isp as nonwall-less versions (Fig. 6.13) [66].

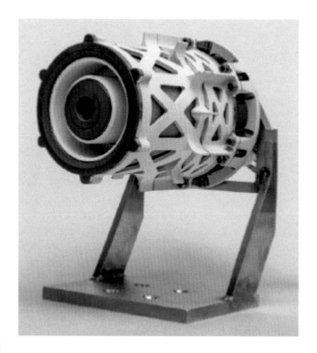

FIGURE 6.12

The engineering model of the MaSMi Hall thruster fabricated by Apollo Fusion, and notional drawing of the magnetic shielding concept [63].

E. Dale, B. Jorns, A. Gallimore, Future directions for electric propulsion research, Aerospace 7 120 (2020).

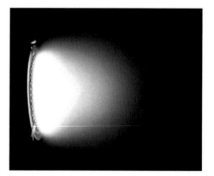

FIGURE 6.13

The ISCT200 in wall-less mode with a gridded anode at the channel exit plane [67].

Left: K. Matyash , R. Schneider, S. Mazouffre , S. Tsikata, L. Grimaud, Rotating spoke instabilities in a wall-less Hall thruster: Simulations, Plasma Sources Sci. Technol. 28 044002 (2019).

6.4.2.3 Electrospray thrusters

JPL developed a microfluidic electrospray propulsion (MEP) thruster with indium propellant. Testing of the thruster demonstrated 120 μN of thrust, >3100 s specific impulse, < 5 W power, at an extraction voltage of 1470 V. The thruster has a dry mass and volume of 26 g and 9 cm^3, respectively [68]. The test article had 400 separate emitters formed with deep reactive ion etching (Fig. 6.14) [69].

The BET-300-P is an electrospray thruster developed by the Busek Company in the United States. It is a successor to the thruster used on the LISA mission, also developed by Busek. The BET-300-P uses nine rows of emitters with a common extraction electrode. The thruster can produce up to 150 μN at a peak power of 2.64 W [70]. The specific impulse ranges from 840 to 1050 s.

The IFM Micro Thruster is a scaled-up version of the IFM Nano Thruster developed by ENPULSION in Austria. It can generate 0.2−1.7 mN and 1000−6000 s at 20−100 W of power [71]. Like the IFM Nano, the Micro can be arrayed to produce more thrust.

Michigan Technological University developed the ionic liquid ferrofluid micro-thruster. The thruster is an electrospray type using an ionic liquid. The unique aspect of this thruster is the use of a ferrofluid and magnetic field to form the needles that the ionic liquid coats. Ferrofluid is a suspension of nanometer-sized ferromagnetic particles such as iron in a viscous solution. Brownian motion keeps the nanoparticles from settling. The ferrofluid responds to external magnetic fields and forms conical peaks. In the Ferrofluid electrospray thruster, the number of peaks can be controlled with an electromagnet. The performance of a single peak was estimated to be 0.38 μN and 1385 s [72].

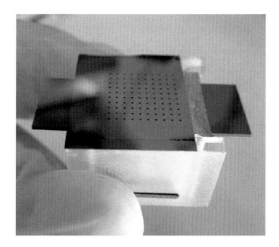

FIGURE 6.14

The JPL MEP thruster [68].

https://www.nasa.gov/sites/default/files/atoms/files/fs_gcdpo_160413_0.pdf.

The Multi-mode Integrated Monopropellant Electrospray (MIME) thruster was developed by the Missouri University of Science and Technology [73], the University of Illinois [74], and commercialized by Froberg Aerospace [75]. The MIME is a dual-mode thruster where a single propellant can be used in electrospray mode or chemical combustion mode. It uses a capillary tube to form the Taylor cone, as opposed to needles. The tube material also acts as a catalyst when heated to allow ignition of the ionic liquid propellant (Fig. 6.15).

The FT-150 FEEP thruster was developed by Alta SpA, now Sitael SpA, located in Italy. The FT-150 is a linear geometry FEEP that uses a blade to form the emitter. The thruster was developed and tested as a potential thruster for the LISA Pathfinder mission [76]. The FT-150 produces $0.1-150$ μN and $3000-4500$ s at a nominal power of 6 W [77].

In-FEEP is an indium liquid metal FEEP thruster developed by the European Space Agency [78,79]. The thruster uses tantalum capillary needles filled with indium liquid metal [80]. A single emitter can produce thrust from 1 to 100 μN and $1600-8000$ s at 13 W of power [78].

The NanoFEEP is a miniaturized FEEP thruster developed by TU Dresden [81,82]. It is being developed for use on the CubeSats UVE-4 and SNUSAT-2. The NanoFEEP uses porous tungsten needles to form the Taylor cones and allow passive propellant feeding via capillary action. Gallium is used as the propellant in the units developed for flight. The overall thruster is very small at less than 6 g. The thruster has a maximum thrust of 22 μN with a specific impulse of up to 6000 s [81]. The power required is on the order of a few watts (Fig. 6.16).

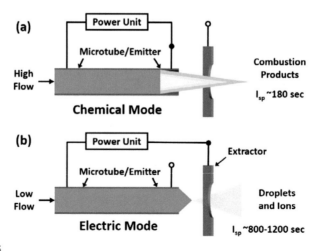

FIGURE 6.15

Description of the MIME thruster from Froberg Aerospace [75].

J. Rovey, C. Lyne, A. Mundahl, N. Rasmont, M. Glascock, M. Wainwright, S. Berg, Review of multimode space propulsion, Prog. Aero. Sci. 118 (2020) 100627.

FIGURE 6.16

The NanoFEEP with a 1 Euro coin for scale and the thruster in testing [81].

Left and Right: D. Bock, M. Tajmar, Highly miniaturized FEEP propulsion system (NanoFEEP) for attitude and orbit control of CubeSats, 144 (2018) 422–428.

    The University of Southampton in the United Kingdom developed a porous emitter electrospray thruster (PET-100). The emitter needles are made of borosilicate glass beads and CNC machined. Performance estimates indicate the thrust of 5−223 μN and a specific impulse of 6000−8500 s at a maximum power of 14 W [83].

6.5 Challenges and future

The challenges for miniature electrostatic propulsion on a commercial scale are missions that require or can tolerate very low thrust but high I_{SP} propulsion. Miniature cold/warm gas or monopropellant thrusters can fill the need for limited ΔV missions typical of most small satellites launched into LEO. These satellites typically have lifetimes of a few months to a few years at best. The ΔV requirement for orbit maintenance may be low enough where a heritage and simple chemical thruster is preferred. Large electrostatic thrusters have been used for Earth-orbiting satellites with 15+ year lifetimes, or deep space exploration missions past Mars. In both cases, the total ΔV requirements drive the propulsion toward high I_{SP} to reduce the propellant mass. The SpaceX Starlink satellites are one example of a niche where miniature electrostatic propulsion may be preferred. Namely, for very low orbit where atmospheric drag is a non-negligible factor where the propulsion systems must constantly provide maintenance thrust against small forces. Small satellites for deep space exploration are another possible avenue, though the radiation-hardened technology for these missions has not yet been proven.

    On the more technical side, the future challenges to miniature electrostatic thrusters are optimization and efficiency. These two are strongly coupled on the system level of the entire spacecraft and must be considered together.

6.5.1 Optimization

Optimization typically refers to performance optimization. For electrostatic thrusters, this can be defined in a number of ways, depending on the desired outcome. Common measures of performance optimization are I_{SP}, thrust-to-power ratio, and overall thruster efficiency. The means to improve these metrics for miniature gridded ion engines and HETs are the same as their larger varieties, with one exception. As both gridded ion engines and HETs create the plasma through "volumetric" ionization, that is, they require the electrons to impact neutrals in the gas, the smaller the thruster size, the less efficient the ionization. The magnetic field needed to confine electrons does not change with thruster size; therefore, electron-bombardment gridded ion engines at small scales can have difficulties in proper cusp magnetic field arrangement to provide effective trapping of the primary electrons from the discharge cathode. Small HETs that use electromagnets to produce the radial magnetic fields can have challenges with too much resistive heating in the magnetic wires that can cause melting of the insulation and magnet failure. For this reason, and the fact that electromagnets require power, many small HETs use permanent magnets.

A second factor to consider in optimization for gridded ion engines and HETs is the surface-to-volume ratio. The ionization rate is a function of volume, but the surface neutralization or loss rate is a function of the discharge chamber/channel surface area. Therefore, as the discharge chamber shrinks, the ratio of the surface area to volume decreases. For example, the surface-to-volume ratio of a cylinder of radius R and length L is

$$\frac{A}{V} = \frac{2\pi R^2 + 2\pi RL}{\pi R^2 L} = \frac{2}{L} + \frac{2}{R}. \tag{6.7}$$

Thus, as L and R, the dimensions of the thruster, decrease, the A/V ratio increases. For the annular HET, the A/V increase is more severe due to the presence of inner and outer walls. The high surface neutralization decreases the ionization efficiency and thus the power efficiency of the thruster.

An additional ionization consideration is the electron source, that is, the cathode. Cathodes with thermionic materials have a minimum temperature required for emission. The heater power required to turn on the cathode is high and can be larger than the power draw of the rest of the thruster. In larger gridded ion engines and HETs, the cathode can self-heat at sufficiently high discharge currents, which means the heater can be turned off and the system becomes more power efficient. However, miniature thrusters rarely draw more than 1 A of discharge current and cannot self-heat the thermionic cathode. The cathode heater power is always required and dominates the electrical power, even though it produces negligible thrust. Cathode lifetime is determined by either erosion of the orifice or loss/damage of the thermionic material. The latter is a concern for small cathodes as they will have smaller emitters and can withstand less sputtering or poisoning before becoming inoperable. For these reasons, and in particular the heater power, some small gridded ion engines and HETs use simple filament cathodes. These are emissive metal filaments similar to those used in incandescent lightbulbs. When heated by passing a current, the filaments emit electrons that can be used for ionization and beam neutralization. The biggest drawback of filament cathodes is a lifetime. The filaments are very thin wires on the order of a millimeter in diameter. The hot filament slowly loses material due to evaporation until a break develops and the filament fails.

The last type of independent electron source is simply an electron or negative ion extracting thruster. A miniature gridded ion engine can extract electrons by changing the grid polarity. A redesign of the grid holes would be desired to maximize electron extraction. In this fashion, a pair of miniature gridded ion engines can function as one complete engine where one extracts ions to produce thrust and the other extracts electrons for beam neutralization. This assumes the plasma in the discharge chamber is generated with RF power or some other method than electron-bombardment and thus does not require a discharge cathode. Such RF neutralizers have been used on the Hayabusa 2 gridded ion engine [84] and RIT-μX [27].

References

[1] NASA, 2018. NASA's Evolutionary Xenon Thruster—Commerical (NEXT—C).
[2] Hofer, R.R., et al., 2015. Development approach and status of the 12.5 kW HERMeS Hall thruster for the solar electric propulsion technology demonstration mission. In: 34th Int. Electr. Propuls. Conf. IEPC-2015-186.
[3] Ziemer, J., et al., July 2005. Colloid Micro-Newton Thruster Development for the ST7-DRS and LISA Missions, pp. 1—9.
[4] Choueiri, E.Y., 2004. A critical history of electric propulsion: the first 50 years (1906—1965). J. Propul. Power 20 (2), 193—203.
[5] Goebel, D.M., Katz, I., 2008. Fundamentals of Electric Propulsion: Ion and Hall Thrusters. Jet Propulsion Laboratory, Pasadena.
[6] Goebel, D.M., Watkins, R.M., Jameson, K.K., 2007. LaB6 hollow cathodes for ion and Hall thrusters. J. Propul. Power 23 (3), 552—558.
[7] Chiu, Y.H., Dressler, R.A., 2007. Ionic liquids for space propulsion. ACS Symp. Ser. 975 (August 2007), 138—160.
[8] Garoz, D., et al., 2007. Taylor cones of ionic liquids from capillary tubes as sources of pure ions: the role of surface tension and electrical conductivity. J. Appl. Phys. 102 (6).
[9] Berg, S.P., Rovey, J.L., 2013. Assessment of imidazole-based ionic liquids as dual-mode spacecraft propellants. J. Propul. Power 29 (2), 339—351.
[10] Szabo, J., et al., 2012. Performance evaluation of an iodine-vapor Hall thruster. J. Propul. Power 28 (4), 848—857.
[11] Szabo, J., Robin, M., Paintal, S., Pote, B., Hruby, V., August 2012. High density Hall thruster propellant investigations. In: 48th AIAA/ASME/SAE/ASEE Joint Propulsion Conference & Exhibit, pp. 1—15.
[12] Dankanich, J.W., Polzin, K.A., Calvert, D., Kamhawi, H., 2014. The iodine satellite (iSAT) Hall thruster demonstration mission concept and development. In: 50th AIAA Joint Propulsion Conference, pp. 1—13.
[13] Polzin, K.A., Peeples, S., 2014. Iodine Hall thruster propellant feed system for a CubeSat. In: 50th AIAA/ASME/SAE/ASEE Jt. Propuls. Conf., pp. 1—10.

[14] Massey, D., Kieckhafer, A.W., Sommerville, J., King, L.B., 2004. Development of a vaporizing liquid bismuth anode for Hall thrusters. In: 40th AIAA Joint Propulsion Conference. AIAA 2004.

[15] Kieckhafer, A., Massey, D., King, L.B., July 2007. Performance and active thermal control of a 2-kW Hall thruster with segmented electrodes. J. Propul. Power 23 (4), 821−827.

[16] Makela, J.M., Washeleski, R.L., Massey, D.R., King, L.B., Hopkins, M.A., 2010. Development of a magnesium and zinc hall-effect thruster. J. Propul. Power 26 (5), 1029−1035.

[17] Jones, R.M., 1984. Comparison of potential electric propulsion systems for orbit transfer. J. Spacecraft Rockets 21 (1), 88−95.

[18] Beattie, J.R., Williams, J.D., Robson, R.R., 1993. Flight qualification of an 18-mN xenon ion thruster. In: 23rd International Electric Propulsion Conference.

[19] Welle, R., 1991. Propellant storage considerations for electric propulsion. In: International Electric Propulsion Conference.

[20] Gartner, W., Zschatzsch, D., Holste, K., Klar, P.J., 2017. Characterization of the operation of RITs with iodine. In: 35th Int. Electr. Propuls. Conf., pp. 1−7.

[21] Saevets, P., Semenenko, D., Albertoni, R., Scremin, G., 2017. Development of a long-life low-power Hall thruster. In: 35th Int. Electr. Propuls. Conf., no. 8−12 October. IEPC-2017-38.

[22] Germany, W., 2011. Vernier propulsion system for small earth remote sensing satellite 'Canopus-V'. System 1−8.

[23] Kawahara, H., Asakawa, J., Yaginuma, K., Koizumi, H., Funase, R., Komurasaki, K., 2015. Ground Experiment for the Small Unified Propulsion System: I-COUPS Installed on the Small Space Probe: PROCYON, pp. 1−13.

[24] Krejci, D., Mier-Hicks, F., Fucetola, C., Lozano, P., Schouten, A.H., Martel, F., 2015. Design and characterization of a scalable ion electrospray propulsion system. In: Jt. Conf. 30th ISTS, 34th IEPC 6th NSAT. Hyogo-Kobe, Japan, pp. 1−11.

[25] Krejci, D., et al., May 2018. Demonstration of the IFM Nano FEEP thruster in low earth orbit. In: 4S Symp. 2018.

[26] Enpulsion, 2019. IFM Product Family [Online]. Available: www.enpulsion.com/order/.

[27] Leiter, H., et al., July 2007. RIT-μx - the new modular high precision micro ion propulsion system. In: 30th Int. Electr. Propuls. Conf., pp. 1−12.

[28] Leiter, H.J., Altmann, C., Porst, J.-P., Lauer, D., Berger, M., Rath, M., 2017. Six decades of thrust - the arian group radiofrequency ion thrusters and systems family. In: 35th Int. Electr. Propuls. Conf. IEPC-2017-027.

[29] Wirz, R., Polk, J., Marrese, C.M., Mueller, J., Escobedo, J., Sheehan, P., 2001. Development and testing of a 3cm electron bombardment micro-ion thruster. In: 27th Int. Electr. Propuls. Conf., pp. 1−12.

[30] Dankongkakul, B., Wirz, R.E., 2016. Design of miniature ring-cusp ion thrusters via analysis of discharge EEDF and plasma parameter mapping. In: 52nd AIAA/SAE/ASEE Jt. Propuls. Conf., pp. 1−14.

[31] Wirz, R.E., 2015. Miniature Ion Thrusters: A Review of Modern Technologies and Mission Capabilities, pp. 1−11.

[32] Tsay, M., Model, J., Barcroft, C., Frongillo, J., Zwahlen, J., Feng, C., 2017. Integrated testing of iodine BIT-3 RF ion propulsion system for 6U CubeSat applications. In: 35th International Electric Propulsion Conference.

[33] Rafalskyi, D., Aanesland, A., 2017. A Neutralizer-Free Gridded Ion Thruster Embedded into A 1U Cubesat Module, pp. 1−8.

[34] Feili, D., Lotz, B., Bonnet, S., Meyer, B.K., Loeb, H.W., August 2009. µNRIT-2 . 5 - a New Optimized Microthruster of Giessen University. IEPC 2009, pp. 1−9.

[35] Koizumi, H., Kuninaka, H., 2011. Performance evaluation of a miniature ion thruster u1 with a unipolar and bipolar operation. In: 32nd International Electric Propulsion Conference.

[36] Trudel, T.A., Bilen, S.G., Micci, M.M., 2009. Design and performance testing of a 1-cm miniature radio-frequency ion thruster. In: 31st International Electric Propulsion Conference, pp. 20−25.

[37] Lubey, D.P., Micci, M.M., Taunay, P., Student, U., Design, E., 2011. Design of the Miniature Microwave-Frequency Ion Thruster, pp. 1−8.

[38] Yamauchi, T., Xu, K.G., 2018. A miniature gridded ion thruster using split ring resonator microplasma. In: 2018 Jt. Propuls. Conf., pp. 1−7.

[39] Collingwood, C.M., Corbett, M.H., Jameson, P., Scientist, E.P., Group, E.P., Manager, P., 2009. The MiDGIT Thruster: Development of a Multi-mode Thruster, pp. 1−13.

[40] Yanhui, J., Tianping, Z., Chenchen, W., Yujun, K., 2017. The Latest Development of Low Power Electric Propulsion for Small Spacecraft, pp. 1−6.

[41] Chan, Y.-A., Herdrich, G., 2017. Characterization of an IEC plasma thruster plume by a nude-type faraday probe. In: 35th Int. Electr. Propuls. Conf., pp. 1−10.

[42] Winter, M., Koch, H., 2017. Inertial electrostatic on finement plasma devices - potential thruster technology for very accurate attitude control systems. In: 35th International Electric Propulsion Conference, pp. 1−17.

[43] Dignani, D., Ducci, C., Cifali, G., Rosseti, P., Andrenucci, M., Alta, 2011. HT-100 Hall thruster characterization tests results. In: 32nd Int. Electr. Propuls. Conf., pp. 1−12. IEPC Paper 2011-191.

[44] ExoTerra, 2019. Halo Hall-Effect Thruster, pp. 1−3.

[45] Raitses, Y., et al., July 2001. Studies of non-conventional configuration closed electron drift thrusters. In: 37th Joint Propulsion Conference and Exhibit.

[46] Smirnov, A., Raitses, Y., Fisch, N.J., 2004. Electron cross-field transport in a low power cylindrical Hall thruster. Phys. Plasmas 11 (11), 4922.

[47] Pigeon, C.E., Orr, N.G., Larouche, B.P., Tarantini, V., Bonin, G., Zee, R.E., 2015. A low power cylindrical Hall thruster for next generation microsatellites. In: 29th Annu. AIAA/USU Conf. Small Satell., pp. 1−9.

[48] Shirasaki, A., Tahara, H., 2007. Operational characteristics and plasma measurements in cylindrical Hall thrusters. J. Appl. Phys. 101 (7), 073307.

[49] Nakano, R., Xu, K.G., 2018. Development of a metronome thrust stand for miniature electric propulsion. In: 2018 Jt. Propuls. Conf., pp. 6−12.

[50] Liang, S., Liu, H., Yu, D., 2019. Effect of permanent magnet configuration on discharge characteristics in cylindrical Hall thrusters. Phys. Lett. A 383.

[51] Polzin, K.A., et al., 2007. Performance of a low-power cylindrical Hall thruster. J. Propul. Power 23 (4), 886−888.

[52] Diamant, K., Pollard, J., Raitses, Y., Fisch, N.J., July 2008. Low power cylindrical Hall thruster performance and plume properties. In: 44th AIAA Joint Propulsion Conference, pp. 1−11.

[53] Exotrail, 2019. Exotrail-Product [Online]. Available: https://exotrail.com/product/.

[54] Henri, D., 2018. Exotrail ExoMG Thruster Ignition [Online]. Available: https://vimeo.com/270366157.

[55] Adamowski, J., September 2018. French space startup raises $4.1 million to develop smallsat electric thruster technology, software. Space News.

[56] Ito, T., Gascon, N., Crawford, W.S., Cappelli, M.A., September 2007. Experimental characterization of a micro-Hall thruster. J. Propul. Power 23 (5), 1068−1074.

[57] Courtney, D.G., Lozano, P., Martinez-Sanchez, M., 2008. Continued investigation of diverging cusped field thruster. In: 44th AIAA Joint Propulsion Conference, vol. AIAA 2008-. Hartford, CT.

[58] Courtney, D.G., Martínez-Sánchez, M., 2007. Diverging cusped-field Hall thruster (DCHT). In: 30th Int. Electr. Propuls. Conf. IEPC-2007-39.

[59] Harle, T., Knoll, A., Lappas, V.J., Bianco, P., 2013. Performance measurements of a high powered Quad confinement thruster. In: 33rd Int. Electr. Propuls. Conf., pp. 1−5.

[60] Fabris, A.L., et al., 2017. Ion Acceleration in a Quad Confinement Thruster, pp. 1−8.

[61] Conversano, R.W., Goebel, D.M., Hofer, R.R., Mikellides, I.G., Katz, I., Wirz, R.E., 2015. Magnetically shielded miniature Hall thruster: design improvement and performance analysis. In: 34th Int. Electr. Propuls. Conf., pp. 1−12. IEPC-2015-100.

[62] Conversano, R.W., Lobbia, R.B., Kerber, T.V., Tilley, K.C., Goebel, D.M., Reilly, S.W., 2019. Performance characterization of a low-power magnetically shielded Hall thruster with an internally-mounted hollow cathode. Plasma Sources Sci. Technol. 28 (10).

[63] Conversano, R.W., Reilly, S.W., V Kerber, T., Brooks, J.W., Goebel, D.M., 2019. Development of and acceptance test preparations for the thruster component of the ascendant sub-kW transcelestial electric propulsion system (ASTRAEUS). In: Proc. 36th IEPC. IEPC-2019-283.

[64] Orbion Space Technology, 2019. Aurora Datasheet.

[65] Sommerville, J., 2019. Performance of the Aurora low-power Hall-effect thruster. In: 36th Int. Electr. Propuls. Conf., pp. 1−7.

[66] Mazouffre, S., Vaudolon, J., Tsikata, S., Harribey, D., Rossi, A., 2015. Optimization of the Design of a Wall-Less Hall thruster. In: 34th Int. Electr. Propuls. Conf., pp. 1−11. IEPC-2015-182.

[67] Matyash, K., Schneider, R., Mazouffre, S., Tsikata, S., Grimaud, L., January 2019. Rotating spoke instabilities in a wall-less Hall thruster: simulations. arXiv.

[68] Marrese-Reading, C., et al., 2016. Microfluidic electrospray propulsion(MEP) thruster performance with microfabricated emitter arrays for indium propellant. In: 52nd AIAA/SAE/ASEE Joint Propulsion Conference.

[69] Rouhi, N., et al., 2015. Fabrication of 3-D silicon microneedles using a single-step DRIE process. J. Microelectromech. Syst. 24 (5), 1409−1414.

[70] Courtney, D.G., Alvarez, N., Demmons, N.R., 2018. Electrospray thrusters for small spacecraft control: pulsed and steady state operation. In: 2018 Jt. Propuls. Conf., pp. 1−15.

[71] Enpulsion, 2018. IFM Micro thruster, pp. 1−2.

[72] Jackson, B.A., King, L.B., 2018. Derived performance metrics and angular efficiency of an ionic liquid ferrofluid electrospray propulsion concept. In: 2018 Jt. Propuls. Conf., pp. 1−10.

[73] Berg, S.P., Rovey, J.L., Prince, B.D., Miller, S.W., Bemish, R.J., 2015. Electrospray of an energetic ionic liquid monopropellant for multi-mode micropropulsion applications. In: 51st AIAA/SAE/ASEE Jt. Propuls. Conf., pp. 1−12.

[74] Wainwright, M.J., Rovey, J.L., Miller, S.W., Prince, B.D., Berg, S.P., 2019. Hydroxylammonium nitrate species in a monopropellant electrospray plume. J. Propul. Power 35 (5), 922–929.

[75] Froberg Aerospace, 2019 [Online]. Available: www.frobergaerospace.com.

[76] Ceccanti, F., et al., 2009. 3200 hours endurance testing of the lisa pathfinder. In: 31st International Electric Propulsion Conference, pp. 1–9.

[77] Paita, L., et al., 2009. Alta's FT-150 FEEP microthruster: development and qualification status. In: 31st Int. Electr. Propuls. Conf., vol. 186, pp. 1–9.

[78] Tajmar, M., Genovese, A., Steiger, W., 2004. Indium field emission electric propulsion microthruster experimental characterization. J. Propul. Power 20 (2), 211–218.

[79] Nicolini, D., Nicolini, D., Chesta, E., Chesta, E., Amo, J.G., Amo, J.G., 2001. Plume characteristics of the indium needle emitter (InFEEP) thruster. Society 15–19.

[80] Tajmar, M., January 2015. Overview of indium LMIS for the NASA-MMS mission and its Suitability for an In-FEEP Thruster on LISA, pp. 1–9.

[81] Bock, D., et al., 2017. In-plume thrust measurement of NanoFEEP thruster with a force measuring probe using laser interferometry. In: 35th Int. Electr. Propuls. Conf., pp. 1–9.

[82] Bock, D., Tajmar, M., 2018. Highly miniaturized FEEP propulsion system (NanoFEEP) for attitude and orbit control of CubeSats. Acta Astronaut. 144, 422–428.

[83] Ma, C., Ryan, C.N., May 2018. The design and characterization of a porous-emitter electrospray thruster (PET-100) for interplanetary CubeSats. In: Proceeding 7th Interplanet. CubeSat Work, pp. 1–9.

[84] Nishiyama, K., Hosoda, S., Ueno, K., Tsukizaki, R., Kuninaka, H., 2016. Development and testing of the Hayabusa2 ion engine system. Trans. Japan Soc. Aeronaut. Space Sci. Aerosp. Technol. Japan 14 (ists30), Pb_131–Pb_140.

Electromagnetic microthrusters

Kristina Lemmer, PhD [1], Gabe Xu, PhD [2]

[1]*Associate Professor, Department of Mechanical and Aerospace Engineering, Western Michigan University, Kalamazoo, MI, United States;* [2]*Associate Professor, Mechanical and Aerospace Engineering, University of Alabama in Huntsville, Huntsville, AL, United States*

7.1 Background

Electromagnetic propulsion systems are a form of electric propulsion that utilize the Lorentz force, F, that results from crossed electric, E, and self-induced and/or externally applied magnetic, B, fields to ionize propellant and create thrust.

$$F = q(E + v \times B) \tag{7.1}$$

where q is the charge state of the particle and v is the instantaneous velocity of the particle within the fields. Electromagnetic thrusters include systems that traditionally operate at large power levels (several to hundreds of kilowatts), including field reversed configuration (FRC) thrusters, magnetoplasmadynamic (MPD) thrusters, pulsed inductive thrusters (PITs), and helicon-based thrusters such as the Variable Specific Impulse Magnetoplasma Rocket (VASIMR). Due to the high power requirements and overall large size and mass of these systems, they are not appropriate for use on small satellites and are not considered a form of microthruster. Other types of electromagnetic thrusters that are appropriate for small satellites include pulsed plasma thrusters (PPTs), micropulsed inductive thrusters (µPITs), and vacuum arc thrusters (VATs). Additionally, micropropulsion systems that utilize magnetic nozzles for acceleration are also included in this category. Fig. 7.1 shows examples of a variety of electromagnetic thrusters. As MPDs, PITs, VASIMR, and FRCs are not applicable to micropropulsion, the reader is encouraged to review papers and texts on the topics if interested [3–9].

While electrostatic and electrothermal propulsion methods were initially discussed and presented at the onset of the age of rocketry in the early 1900s, electromagnetic propulsion concepts did not begin to emerge until the latter half of the 20th century [10]. Electromagnetic systems underwent a period of intense development in the 1960s and 1970s, and as a result, the first successful operation of an EP system in space was that of a PPT for attitude control on board the Russian ZOND-2 spacecraft in 1964 [3,11]. MPDs, PPTs, and VATs have flight heritage; however, only PPTs and VATs have been used on operational spacecraft while MPDs have been flown on flight demonstration missions, most of which occurred on suborbital trajectories [3,12–14].

FIGURE 7.1

(Left) Plume of an MPD thruster; (middle) Electrodeless Lorentz Force (ELF) thruster [1]; (right) Busek pulsed plasma thruster [2].

Left: credit: NASA.

For small satellites, one of the earliest types of EP systems adopted was a form of electromagnetic microthruster. Solid propellant PPTs and VATs have been gaining popularity for their compatibility with small satellites and the restrictions that accompany being launched as a secondary payload such as the amount of stored energy a satellite is allowed, risks of carrying corrosive propellant, and compressed gas limitations. PPTs can use inert, solid propellants, so there are no issues with stored energy, corrosive liquids, or compressed gasses. They are pulsed systems, so PPTs and VATs can operate to the power limitations of a small satellite by adjusting the rate at which the thruster is pulsed. A 3U CubeSat without deployable solar panels is typically limited to a maximum power of about 15 W. The first magnetic nozzle microthruster was launched in January 2021, the Maxwell by Phase Four, but no on-orbit data are yet available. To date, no uPITs have flow; however, there is extensive ongoing research into the technology.

7.2 Thruster types

While all electromagnetic systems rely on the basic principle of crossed electric and magnetic fields to accelerate a plasma, different thruster types achieve performance goals using very different concepts of operation. Electrostatic thrusters all follow the principles of ionization, acceleration, and neutralization; however, electromagnetic thrusters generally combine ionization and acceleration into one process and do not require a neutralization stage. The plasma produced by electromagnetic thrusters is assumed to be quasi-neutral, meaning that there is the same number of electrons and ions. As the propellant acceleration mechanism is due to the Lorentz force, both the ions and electrons are accelerated. This differs from electrostatic thrusters that rely on a potential difference to accelerate ions only. The following sections will differ from the previous chapter on electrostatic micropropulsion as the principles of operation are quite different between thruster types. Rather than discuss the different operation mechanisms, specific thruster types will be discussed.

7.2.1 Pulsed plasma thrusters and vacuum arc thrusters

Both solid propellant PPTs and VATs operate with similar principles. A solid propellant (usually polytetrafluoroethylene (PTFE) in the case of a PPT, and a conductive metal in the case of a VAT) is spring fed. A small spark initiates a discharge and ablates some of the propellant, causing a path for an arc to form between an anode and a cathode. This arc ionizes the ablated propellant. Energy from the strong instantaneous current generated by the arc is deposited into a self-induced magnetic field that is orthogonal to the current. This results in the Lorentz force accelerating the quasi-neutral plasma downstream, as shown in Fig. 7.2. For the case of the VAT, the propellant is the cathode material as current is conducted through the vapor sublimated from the cathode surface. The anode is typically an electrode surrounding the cathode propellant with an insulator between the two, as shown in Fig. 7.2 [15]. Solid propellant PPTs and VATs have efficiencies that range from 8% to 55% [3]. The wide range of efficiencies is due to the fact that a substantial portion of ablated propellant may be neutral species that are accelerated gas dynamically rather than electromagnetically. This is dependent on the thruster design and operating conditions. Both metallic and fluoride-containing neutral particles have been found in the exhaust of PPTs [16]. The metallic particles are likely from electrode erosion. One hypothesis on the presence of fluoride-containing neutral particles states that UV radiation produced by the plasma is absorbed by PTFE propellant. This causes a high-pressure vapor generation beneath the surface of the PTFE. Another source of neutral particles comes from late-cycle evaporation of the PTFE propellant as a result of heating from the plasma discharge [17,18].

PPT and VAT performance measurements from a variety of thruster geometries and circuit designs have resulted in empirical models of performance trends. In general, impulse bit (the total impulse delivered per thruster discharge) and specific impulse (Isp) both increase with increasing capacitor energy. Thruster performance can be modified by changing the separation distance between the capacitor electrodes as well as changing the feed method and direction of the solid propellant in a PPT. One example of this is the surface area of the propellant exposed to the discharge. When

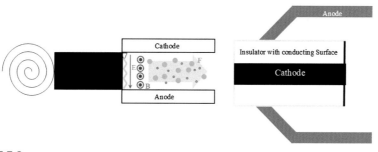

FIGURE 7.2

Left: Principle of operation for pulsed plasma thrusters and vacuum arc thrusters. Right: Schematic of electrode configuration in vacuum arc thrusters.

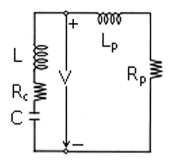

FIGURE 7.3

Equivalent circuit for PPT. L is capacitor inductance, R_c is capacitor resistance, C is capacitance, L_p is the time-varying inductance of the arc, and R_p is the time-varying resistance of the plasma.

the surface area of the propellant is larger for a given input energy, more propellant is ablated, causing the impulse bit to increase and the specific impulse to decrease [19]. Note that first principles analysis of PPTs is lacking mostly due to the fact that, for example, the quantitative relationship between the magnitude of the impulse bit and the amount of stored energy in the capacitor is very sensitive to initial plasma conditions and field fringing effects at the walls of the accelerator.

Acceleration in PPTs is typically modeled using circuit analysis where the rapidly moving arc is counted as a variable inductance. Fig. 7.3 shows a schematic of the equivalent circuit of a PPT. With this model, standard circuit analysis techniques can be used to determine the position of the arc as a function of time. A rapid change in inductance occurs as the arc traverses downstream, resulting in an induced voltage drop due to the dynamic nature of the discharge. The largest hurdle in modeing the acceleration process in PPTs and VATs comes from knowing the plasma properties that predict the time-varying resistance of the plasma, R_p, and the time-varying inductance of the arc, L_p [3].

7.2.1.1 Alternative PPT/VAT technologies

Solid propellant PPTs typically suffer from late ablation and the presence of thermally ejected macroparticles. This results in a low-efficiency limit of about 15% [17]. Additional issues include variability in the impulse bit, low mass flow control, and propellant contamination. Therefore, PPTs have been developed and tested that operate on liquid and gaseous propellants. The benefits and drawbacks of these systems result from the introduction of a propellant feed system. While the feed system significantly complicates the thruster design, it can also limit the amount of liquid or gas that is available for ionization and acceleration. Thus, fine control over the impulse bit is achievable, and there are significant efficiency gains as mass flow is shut off between pulses. Furthermore, the risk of contamination is lower due to using inert gases/liquids. Some liquid propellants that have been tested include perfluoropolyether, alcohol, and dimethyl ether [20]. Gas-fed PPTs offer even more efficiency and impulse bit control; however, they also add more complexity with the need to store a compressed gas. Gasses that have been tested include hydrogen, nitrogen, argon, and xenon.

PPTs and VATS have been designed with applied external magnetic fields in an effort to increase efficiency by enhancing acceleration and reducing plume divergence [3,21]. The applied magnetic field has also been shown to decrease the impulse bit, likely as a result of an increase in the acceleration field causing the residence time of the arc near the propellant surface to decrease [22].

7.2.1.2 Lifetime and flight considerations

For low thrust situations with discharge energies in the range of a few joules, thruster lifetime is dependent on the amount of propellant available. No significant life-limiting factors exist in these cases. However, when the energy per pulse is increased to several hundred joules, erosion of the anode becomes a severe issue [23]. Anode melting and erosion can be minimized by reducing the residence time of the arc near the exit plane. This can be done by increasing the length of the electrodes. Increasing discharge energy while keeping the electrode length the same causes the arc to reach the end of the electrodes without the capacitor fully discharging, thus increasing the residence time of the arc at the exit plane [24,25]. As a result, PPT and VAT design considerations must be made to ensure that the arc does not traverse the length of the electrodes before the capacitor is fully discharged.

While anode erosion must be considered for spacecraft contamination issues, the most serious problem with PPT and VAT systems comes from electromagnetic interference (EMI) induced by the high voltage and current transients that exist during the thruster discharge. In the case of a CubeSat, where the spacecraft is severely volume limited, EMI may not be avoidable, and necessary shielding techniques must be implemented.

7.2.2 Magnetic nozzle thrusters

Magnetic nozzle thrusters are a category that combines an efficient ionization process with expansion through a magnetic nozzle. The first magnetic nozzle-type thruster was flown in January 2021. The Maxwell thruster was designed and built by PhaseFour; however, with operating powers of 300−500 W, it lies outside the range of what is considered micropropulsion. Extensive research is underway to create smaller magentic nozzle thrusters. Ionization processes under development include electron cyclotron resonance heating, inductively coupled plasma formation, and helicon plasma formation. These ionization methods are all forms of electromagnetic (EM) waves. They are electrodeless and utilize alternating current electric fields in the radio frequency and microwave range. The use of EM waves for ionization avoids a common life-limiting factor of many types of EP systems. While magnetic nozzles are not the most efficient method for acceleration of a plasma, interest remains because magnetic nozzles do not have erosion issues that are observed in all currently flying microelectric propulsion devices (i.e., electrospray, gridded ion engine, Hall effect thruster, PPT, and

VAT). This also means that these devices do not have the propellant limitations of most electrostatic thruster technologies. As a result, there is significant interest in magnetic nozzle thrusters for in-situ resource utilization technologies. Systems under development include those that can operate on water, methane, and other propellants available in the environments of interest for exploration.

Electron cyclotron resonance (ECR) occurs when the frequency of incident radiation coincides with the rotation frequency of electrons about a magnetic field. ECR is a very efficient method for ionization utilizing resonance between the electron cyclotron frequency, ω_{ce} (Eq. 7.2) and the incident frequency of a microwave source.

$$\omega_{ce} = \frac{eB}{m_e} \tag{7.2}$$

where e is the elementary charge, B is the magnetic field strength, and m_e is the mass of an electron. For a microwave frequency of 2.45 GHz, $B = 875$ G. As the alternating electric field of the microwaves is synchronous with the gyration period of free electrons, the kinetic energy of the free electrons increases. When those electrons are incident on a neutral particle, they can cause ionization. Fig. 7.4 shows a schematic of an ECR thruster with a magnetic nozzle.

Alternative EM methods for ionization within magnetic nozzle thrusters include inductively coupled plasma sources and helicon plasma sources (shown in Fig. 7.5). Both utilize alternating current electric fields at radio frequencies. Helicon plasma sources additionally have an externally applied magnetic field which causes the electric field to be nearly perpendicular to the electric current. Both helicon and ECR sources are more efficient sources of plasma generation when compared with an inductively coupled source; however, the complexity and power required to add an external magnetic field diminishes the benefits for a microsatellite.

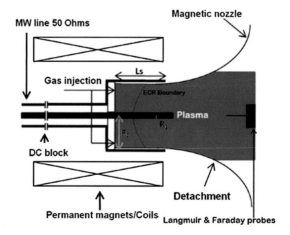

FIGURE 7.4

Schematic of a microwave-powered electron cyclotron resonance thruster with a permanent magnet nozzle [26].

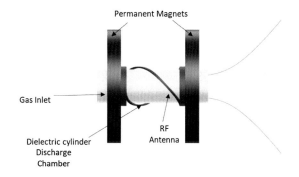

Permanent Magnets

Gas Inlet

RF
Antenna

Dielectric cylinder
Discharge
Chamber

FIGURE 7.5

Schematic of a magnetic nozzle thruster with helicon ionization.

7.3 Current state of the art

As discussed previously, there are few electromagnetic micropropulsion systems with flight heritage. In fact, the only systems with any flight heritage are PPTs, VATs. No system utilizing a magnetic nozzle for acceleration has been launched for use on small spacecraft. In fact, the first magnetic nozzle system was launched in January 2021. As previousy discussed, this systems requires power-levels that place it outside the scope for small satellites. Regardless, there are several types of magnetic nozzle systems that are currently under development.

7.3.1 Systems with flight heritage

The first EP systems to fly in the United States were PPTs, leading to their early development. Furthermore, the simplicity and solid propellant of PPTs and VATs have led them to be among the most implemented propulsion source on microsatellites. It should be noted that most of the systems that have been launched were developed by universities and not mass-produced. Rather, the thrusters were developed specifically for the assigned mission. Assuming a small satellite is one under 50 kg, the following systems have flight heritage.

In 2007, FalconSat-III was launched with PPTs from Busek Company, Inc. for technology demonstration and attitude control. Fig. 7.6 shows FalconSat-III, which was a 50-kg satellite that was designed, built, and operated by the Air Force Academy. The PPTs operated successfully, demonstrating the ability to control the spacecraft's attitude.

The Ballistic Reinforced Communications Satellite (BRICSat-P), Fig. 7.7, is a 1.5U CubeSat built by the US Naval Academy that demonstrated the use of VATs onboard for satellite detumbling. The Micro-Cathode Arc Thruster (μCAT) designed by researchers at George Washington University (GWU) was first tested on this satellite. Four thrusters were placed around the center of gravity and successfully performed the detumbling mission [28]. The thruster technology has since been expanded to include a magnetic field enhancement and was supposed to be used

FIGURE 7.6

FalconSat-III (left) launched with PPTs (right—Busek website) for attitude control.

FIGURE 7.7

BRICSat-P (left) launched with μCAT VATs from George Washington University (middle) for detumbling. μCAT firing in the vacuum facility at GWU (right) [27].

for formation flying and to maintain orbit on the South Korean CANYVAL-X satellite, a 2.7 kg 2U CubeSat. Unfortunately, the satellite failed to operate in space. BRICSat-2 was launched on June 25, 2019, and is currently operational in orbit. Four μCAT thrusters from GWU with an updated externally applied magnetic field are onboard for attitude control and orbit maintenance. To date, no on-orbit data are publicly available.

In 2017, the AOBA VELOX-III satellite, jointly developed by Nanyang Technological University (NTU) in Singapore and Kyushu Institute of Technology in Japan, was successfully deployed from the International Space Station. The satellite carried a dual-axis PPT developed at NTU that was successfully demonstrated to increase mission duration by maintaining orbit. The PPT was also used for attitude control [29,30]. Fig. 7.8 shows a photograph of the PPT used on AOBA VELOX-III.

A sulfur-propellant PPT developed at the University of Washington was deployed on January 31, 2020 onboard HuskySat-1, a 3U CubeSat. Sulfur was used as the propellant to provide twice the specific thrust than traditional PTFE. To date, no data from orbit have been publicly released. The PPT is shown in Fig. 7.9.

FIGURE 7.8

NTU developed dual-axis PPT used on AOBA VELOX-III satellite [30].

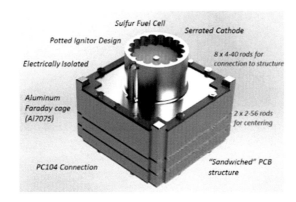

FIGURE 7.9

The University of Washington developed sulfur PPT.

7.3.2 Systems under development

A quick search on the internet for PPTs and VATs will show several options that are currently available for small spacecraft. In fact, numerous PPTs and VATs have been launched, but the satellite failed in some way or other on orbit. Several examples are shown and discussed with interesting advancements in technology. Some thruster concepts utilizing a magnetic nozzle for acceleration with electrodeless ionization methods are also discussed; however, these systems are further from on-orbit demonstration than PPTs and VATs.

NTU has developed a 4-head PPT scheduled to fly aboard the AOBA VELOX-IV satellite based on the experience with their dual-axis PPT on the AOBA VELOX-III satellite. The mission is to demonstrate technologies applicable to lunar study. The PPTs will be used for orbit maintenance and reaction wheel desaturation [29].

Other PPT systems include the PPT for CubeSats developed by Clyde Space and Mars Space Ltd. that is capable of 650 s Isp with impulse bits up to 45 µN-s operating on 2 W power and Fotec's PPT module capable of 900 s Isp and impulse bits ranging from 3.5 to 10 μN-s [31−34]. Both systems operate on solid PTFE propellant.

As previously mentioned, liquid and gas-fed PPTs are of interest for precisely controlled impulse bits and higher efficiencies. One example of a system under development is the Liquid Micro Pulsed Plasma Thruster (LμPPT) from QuinteScience and the Institute of Physics and Laser Microfusion in Poland. The thruster operates on liquid perfluorpolyether, a nonhazardous liquid with properties similar to those of PTFE. In laboratory testing, the thruster has produced impulse bits that range from 10 to 25 mN-s and an Isp ranging from 1000 to 1400 s [35]. Another prototype liquid propellant PPT has been designed and tested by Digital Solid State Propulsion using an in-house developed hydroxylammonium nitrate-based monopropellant (GEM). The unit demonstrated 100 pulses with an average impulse bit of 0.3 mNs and an estimated specific impulse of 214−384 s. This device was designed with the idea of a multimode propulsion system that can operate in both a chemical propellant mode and an electric propulsion mode.

Researchers at George Washington University are improving on the performance from their μCAT VATs through the use of an acceleration electrode. The magneto-plasmadynamic two-stage μCAT improves thrust by almost a factor of two over a traditional VAT while increasing the thrust-to-power ratio [36]. Fig. 7.10 shows a schematic and sketch of the μCAT-MPD.

Past research has worked toward the development of microthrusters utilizing helicon source ionization; however, the miniaturization of the helicon source has yet to be demonstrated successfully. As such, only ECR thruster concepts will be discussed. A research team at Onera, the French aeronautics, space, and defense research laboratory has developed an ECR thruster and published its methods for design [26]. As a result, active interest in ECR thrusters is underway at several institutions. Fig. 7.11 shows an ECR thruster concept operating at the University of Michigan.

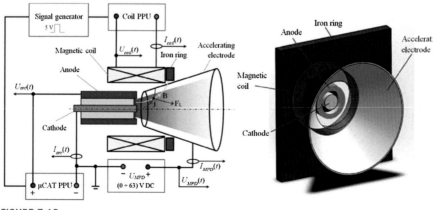

FIGURE 7.10

Acceleration electrode, magnetically enhanced μCAT-MPD [36].

FIGURE 7.11

ECR thruster operating at the University of Michigan.

7.4 Challenges and future

The challenges for miniature electromagnetic propulsion are twofold. Similar to electrostatic propulsion methods, small satellite missions that require or can tolerate very low thrust but high I_{SP} propulsion tends to be limited. Most small satellites launched into LEO have lifetimes of a few months to a maximum of a few years. The ΔV requirements for maintaining orbit for missions with such short lifetimes are low and do not require the use of a highly efficient EP system. Thus, a heritage and simple cold gas or chemical thruster is preferred. An area where small satellites may see a significant use for electromagnetic propulsion is deep space exploration. EM ionization with magnetic nozzles allows for a variety of propellants to be used without the worry of oxidization or corrosion of an electrode. Thus, in-situ propellant utilization can be realized with these types of thrusters.

For electromagnetic propulsion systems with magnetic nozzles, scaling down becomes an issue. The smaller the magnetic nozzle, the stronger the magnetic field required to contain and accelerate the plasma. This has been accomplished through the use of permanent magnets to reduce required power; however, the properties of permanent magnets change when thermal loads are exceeded. Thermal limits on permanent magnets are on the order of thermal loads that are seen in EM excitation methods.

Many of the optimization issues discussed in the electrostatic propulsion chapter are not applicable to electromagnetic propulsion systems. The ion impact ionization method with its volume-limited restrictions is mostly avoided in electromagnetic thrusters such as PPTs, VATs, and magnetic nozzle thrusters. Furthermore, as the Lorentz force is agnostic to the charge of a particle, a neutralizer source is not required for thruster operation. This keeps both the complexity of the thruster system, and the power required to operate a thruster, low. However, the miniaturization of power electronics for RF and microwave systems is complicated. Ongoing research is working to advance power systems for small spacecraft propulsion systems.

References

[1] Kirtley, D., Pancotti, A., Slough, J., Pihl, C., 2012. Steady operation of an FRC thruster on martian atmosphere and liquid water propellants. In: 48the AIAA/ASME/SAE/ASEE Joint Propulsion Conference and Exhibit. AIAA, pp. 2012–4071.

[2] Legge, R.S., Clements, E.B., Shabshelowitz, A., 2017. Enabling microsatellite maneuverability: a survey of microsatellite propulsion technologies. In: 2017 IEEE MTT-S International Microwave Symposium (IMS), pp. 229–232.

[3] Myers, R.M., 1993. Electromagnetic propulsion for spacecraft. In: Aerospace Design Conference. AIAA, pp. 93–1086.

[4] Jahn, R.G., 1996. Physics of Electric Propulsion. Dover Publications, Inc.

[5] Slough, J., Kirtley, D., Weber, T., 2009. Pulsed plasmoid propulsion: the elf thruster. In: 31st International Electric Propulsion Conference, pp. 1–24.

[6] Niemela, C., Kirtley, D., 2008. Initial results on an annular field reversed configuration plasma translation experiment. In: 55th JANNAF Propulsion Meeting/4th Liquid Propulsion Subcommittee/3rd Spacecraft Propulsion Subcommittee/6th Modeling and Simulation Subcommittee Meeting, vol. 12, p. 704.

[7] Polzin, K.A., 2011. Comprehensive review of planar pulsed inductive plasma thruster research and technology. J. Propul. Power 27 (3), 513–531.

[8] Díaz, F.R.C., 2000. The Vasimr rocket. Sci. Am. 283, 90–97.

[9] Cassady, L., et al., 2010. VASIMR performance results. In: 46th AIAA/ASME/SAE/ASEE Joint Propulsion Conference & Exhibit.

[10] Choueiri, E.Y., 2004. A critical history of electric propulsion: the first 50 years (1906–1965). J. Propul. Power 20 (2), 193–203.

[11] Nickolay, N.A., Popov, G.A., Kazeev, M.N., 2013. Ablative pulsed plasma thrusters R&D in Russia since the beginning of the 90s. In: 33rd International Electric Propulsion Conference. IEPC, pp. 2013–2068.

[12] Gorshkov, O.A., Shutov, V.N., Kozubsky, K.N., Ostrovsky, V.G., Obukhov, V.A., 2007. Development of high power magnetoplasmadynamic thrusters in the USSR. In: 30th International Electric Propulsion Conference. IEPC, pp. 2007–2136.

[13] Lemmer, K., 2017. Propulsion for CubeSats. Acta Astronaut. 134, 231–243.

[14] Lev, D., Myers, R.M., Lemmer, K.M., Kolbeck, J., Koizumi, H., Polzin, K., 2019. The technological and commercial expansion of electric propulsion. Acta Astronaut. 159, 213–227.

[15] Polk, J.E., Sekerak, M.J., Ziemer, J.K., Schein, J., Qi, N., Anders, A., 2008. A theoretical analysis of vacuum arc thruster and vacuum arc ion thruster performance. IEEE Trans. Plasma Sci. 36 (5), 2167–2179.

[16] Spanjers, G., Lotspeich, J., McFall, K., Spores, R., 1998. Propellant losses because of particle emission in a pulsed plasma thruster. J. Propul. Power 14 (4), 554–559.

[17] Spanjers, G., McFall, K., Gulczinski III, F., Spores, R., 1996. Investigation of propellant inefficiencies in a pulsed plasma thruster. In: 32nd Joint Propulsion Conference.

[18] Spanjers, G., Malak, J., Leiweke, R., Spores, R., 1998. Effect of propellant temperature on efficiency in the pulsed plasma thruster. J. Propul. Power 14 (4), 545–553.

[19] Palumbo, D.J., Guman, W.J., 1975. AIAA Paper. Effects of Propellant and Electrode Geometry on Pulsed Ablative Plasma Thruster Performance, vols. 75–409.

[20] Sasha, W., Miller, C.S., Ingersoll, J.E., Yost, B.D., Agasid, E., Burton, R., Carlino, R., et al., 2018. State of the Art: Small Spacecraft Technology. NASA Technical Publication. Document ID: 20200001421.

[21] Keidar, M., et al., 2005. Magnetically enhanced vacuum arc thruster. Plasma Sources Sci. Technol. 14 (4), 661–669.

[22] Takegahara, H., Ohtsuka, T., Kimura, I., 1984. Effect of applied magnetic fields on a solid-propellant pulsed plasma thruster. In: International Electric Propulsion Conference. IEPC 84-50.

[23] Palumbo, D.J., 1979. Solid Propellant Pulsed Plasma Propulsion System Development for N-S Station Keeping, vols. 79−2097. AIAA Paper.

[24] Vondra, R.J., 1979. US Air Force Programs in Electric Propulsion, vols. 79−2123. AIAA Paper.

[25] Eckbreth, A.C., 1968. Current Pattern and Gas Flow Stabilization in Pulsed Plasma Accelerators. Princeton University.

[26] Cannat, F., Lafleur, T., Jarrige, J., Chabert, P., Elias, P.Q., Packan, D., 2015. Optimization of a coaxial electron cyclotron resonance plasma thruster with an analytical model. Phys. Plasmas 22 (5).

[27] Keidar, M., 2016. Micro-cathode arc thruster for small satellite propulsion. In: 2016 IEEE Aerospace Conference, pp. 1−7.

[28] Kolbeck, J., et al., 2016. μCAT micro-propulsion solution for autonomous mobile on-orbit diagnostic system. In: 30th Annual AIAA/USU Conference on Small Satellites. SSC16-V-7.

[29] Bui, T.D.V., et al., 2018. Design and development of AOBA VELOX-IV nanosatellite for future lunar horizon glow mission. In: 32nd Annual AIAA/USU Conference on Small Satellites. SSC18-WKIX-02.

[30] Tran, Q.-V., Lim, W.-S., Bui, T.-D.-V., 2019. Kay-soon low, and Bingyin Kang, "development of a dual-axis pulsed plasma thruster for nanosatellite applications. J. Small Satell. 8 (2), 837−847.

[31] Guarducci, F., Coletti, M., 2011. Design and testing of a micro pulsed plasma thruster for cubesat applicaion. In: 32nd International Electric Propulsion Conference. IEPC, pp. 2011−2239.

[32] Coletti, M., Guarducci, F., Gabriel, S.B., 2011. A micro PPT for Cubesat application: design and preliminary experimental results. Acta Astronaut. 69 (3−4), 200−208.

[33] Krejci, D., Mier-Hicks, F., Fucetola, C., Lozano, P., Schouten, A.H., Martel, F., 2015. Design and characterization of a scalable ion electrospray propulsion system. In: 34th International Electric Propulsion Conference. IEPC, pp. 2015−2149.

[34] Krejci, D., Seifert, B., Scharlemann, C., 2013. Endurance testing of a pulsed plasma thruster for nanosatellites. Acta Astronaut. 91, 187−193.

[35] Barral, S., et al., 2015. Time-of-flight spectrometry and performance of a pulsed plasma thruster with non-volatile propellant. In: 34th International Electric Propulsion Conference.

[36] Zolotukhin, D.B., Daniels, K.P., Bandaru, S.R.P., Keidar, M., 2019. Magnetoplasmadynamic two-stage micro-cathode arc thruster for CubeSats. Plasma Sources Sci. Technol. 28 (10), 105001.

Further reading

[1] Nishiyama, K., Hosoda, S., Ueno, K., Tsukizaki, R., Kuninaka, H., 2016. Development and testing of the Hayabusa2 ion engine system. Trans. Japan Soc. Aeronaut. Space Sci. Aerosp. Technol. Japan 14 (ists30), Pb_131−Pb_140.

[2] Leiter, H., et al., 2007. RIT-μX - the new modular high precision micro ion propulsion system. In: 30th Int. Electr. Propuls. Conf., pp. 1−12.

Related development

Thrust measurement

Akira Kakami, PhD

Associate Professor, Department of Aeronautics and Astronautics, Tokyo Metropolitan University, Hino, Tokyo, Japan

8.1 Thrust stand

Space propulsion devices such as electric and chemical propulsion, yield a wide-range thrust from nN to kN and have weights from grams to ten of kg. Hence, some thrusters have a thrust-to-weight ratio below 1. Especially for low thrust-to-weight ratios, the thrust was measured with thrust stands. Generally, the pendulum type is used since the thrust direction is perpendicular to the gravity to cancel the negative influence of the thruster weight. The chapter introduces the displacement and null-balance methods and explains the theory and dynamic properties.

This chapter also addresses the thrust-target method, which is an alternative when the thrust stand is not applicable due to thrust level or thruster weight. The thrust target collides with the exhaust jet, and the thrust and impulse are evaluated from the amplitude of the pendulum-arm displacement. Further, calibration methods will be explained, including the theory and examples.

8.1.1 Introduction

The thrust stand, which is sometimes referred to as the thrust balance, is the device that measures the thrust of space propulsion devices. Fig. 8.1 illustrates a typical horizontal pendulum type. Such type has been developed since the dawn of space age [1−6], and the established fundamental design was found in Fig. 8.2, which is one of the schematics in a 1971 patent [7]. The reference [8] reports the thrust stand technology related to the patent. The use of the pendulum makes thrust direction perpendicular to the gravity so that the thruster weight has less influence on the measurement. The steady and impulsive thrusts are evaluable using the displacement and displacement amplitude of the pendulum arm, respectively. We can understand the principle using high-school physics; the displacement and amplitude are proportional to the external force and impulsive force, respectively. Despite the straightforward principle, in design, analysis based on the vibration theory is necessary due to extra elements such as a damper. A damper is usually used to suppress the oscillation. Without a damper, the pendulum arm keeps oscillating in the vacuum because the torsional hinge produces negligible friction. Even in the atmospheric

Space Micropropulsion for Nanosatellites. https://doi.org/10.1016/B978-0-12-819037-1.00010-4

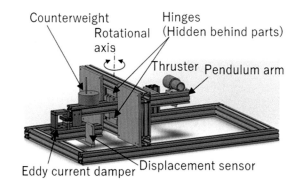

FIGURE 8.1

A thrust stand.

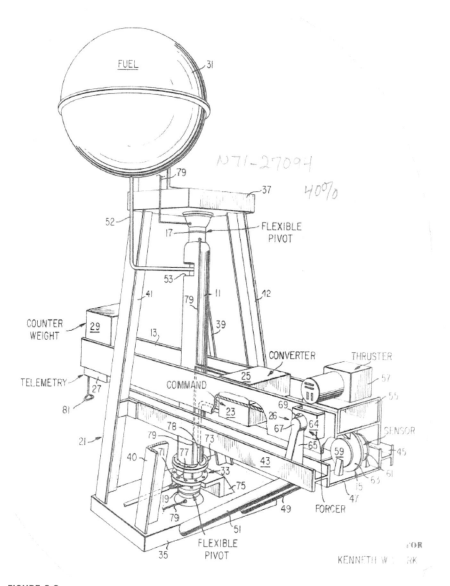

FIGURE 8.2

A pendulum-type thrust stand on patent [7].

environment, the oscillation is usually maintained because the air yields a weak damping force. The counterweights are sometimes placed on the end opposing to the thruster to reduce mechanical noise. Even in the quiet laboratory, vacuum pumps and human walk induce vibration, which is transferred into the pendulum arms. The external vibration causes the pendulum-arm oscillation when the center of gravity (C.G.) of the pendulum arm is not placed on the rotational axis. However, the settlement of C.G. on the rotational axis can eliminate the mechanical noise transfer. Hence, the counterweights are occasionally used, especially for low thrust-to-weight thrusters.

The null-balance method, the principle of balance scales, is also applied to performance evaluation. In the classical balance scales, a mass to be evaluated is put on a beam side, and then known-mass weights are stacked on the other side so that the beam becomes horizontal. The mass is determined from the total mass of the settled weights and the distance from the fulcrum. In thrust measurement, stacking and selecting mass is conducted by a controller, and actuators are used instead of the known masses. The method is referred to as the null-balance method, with which the nonlinear displacement sensor with high resolution is applicable. The application of the null-balance method to thrust stands was also patented [7] and reported [8]. The method and theory will be discussed in Section 8.1.3.

In high-power rocket engines for launching and sounding rockets, thrust is usually evaluated using load cells where strain gauges or piezoelectric/resistive gauges measure the material distortion induced by thrusts. The load cells have various features: simpleness, compact, and low price. The stiffness, which directly affects the dynamic response, is readily augmented by the use of high elastic modulus materials. Fig. 8.3 presents an example of the thrust stand for rocket engines, the thrust of which is much larger than the weight [9]. Aside from these rockets, in some comparatively large thrust thrusters for spacecraft, the performance has been evaluated using the load cells either.

The R-4D bipropellant thruster, which was developed for the Apollo Project and was installed on the Luna and Service Modules for the reaction control, was prototyped and tested in the 1960s. The thruster weighted 3.6 kgf and produced 490-N thrust with a specific impulse of 290 s (at that time), so that the thrust to weight ratio is approximately equal to 10. The thrust stand shown in Fig. 8.4 evaluated the

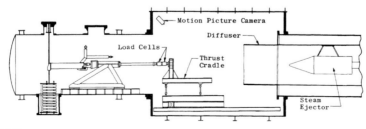

FIGURE 8.3

An example of a load cell—based thrust stand for higher thrust rocket engines [9].

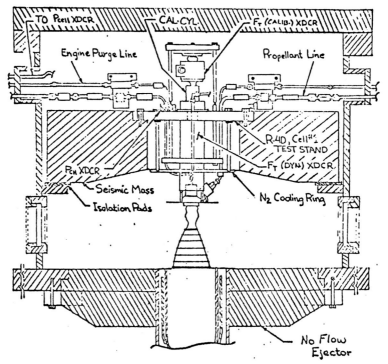

FIGURE 8.4

A thrust stand for R-4D bipropellant thruster [10].

performance [10] and was vertically suspended by a seismic mass. A high stiffness semiconductor load cell measured steady and impulsive thrusts. The thrust stand was periodically calibrated with secondary standard load cells.

Some may consider that the simple miniaturization of such thrust stand allows the performance evaluation of small thrust-to-weight ratio thrusters, which are typical for onboard propulsion for spacecraft. However, the presence of gravity never permits the straightforward approach. For relatively low thrust, lower capacity load cells are mandatory since the linearity or resolution is usually as high as 0.1% full scale. Since the absolute maximum force of the load cell is strongly dependent on the measurable range, some mechanisms are necessary to eliminate the negative effect of gravity. Hence, the pendulum has been introduced to the performance evaluation of onboard propulsion.

8.1.2 The displacement method

Both vertical- and horizontal-type thrust stands are used to determine the thrust of various space propulsion devices [4,8,11]. Fig. 8.5 shows a schematic of the horizontal pendulum type using the displacement method. From the viewpoint of dynamics,

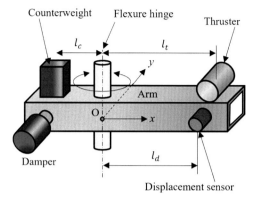

FIGURE 8.5

A dynamical model of the displacement method thrust stand.

the vertical pendulum type is almost identical to the horizontal pendulum type, except that the restoring force originating from gravity needs to be considered. Some groups suspend the thruster using wires, and the wire-suspended types can be categorized into the vertical type. The typical thrust stand has an arm, flexure hinge, counterweight, and displacement sensor and is theoretically identical to the horizontal or ballistic pendulum that oscillates around the flexure hinge. Though the counterweight is not always necessitated, it can suppress the mechanical noise due to the external vibration originating from the vacuum chamber vibration, as will be mentioned later. The section addresses the principles of the displacement-method thrust stand with the frequency responses.

8.1.2.1 Steady thrust

The displacement method determines thrust from the pendulum displacement. Both the horizontal or vertical pendulums yield displacement proportional to the external forces as long as the angular displacement is so tiny that $\sin\theta \approx \theta$ and $\cos\theta \approx 1$, and both pendulums are theoretically the same. The angular equation of motion for the model illustrated in Fig. 8.5 is expressed as:

$$I\ddot{\theta} + c\dot{\theta} + k\theta = l_t T \tag{8.1}$$

$$y_d = l_d\theta \tag{8.2}$$

where θ is the deflection angle of the pendulum arm, I is the moment of inertia, c is the damping coefficient, k is the spring constant, and y_d is the displacement that is evaluated with the displacement sensor. Note that k is the rotational spring constant of the flexure hinges for the horizontal pendulum and the sum of the spring constant and gravity for the vertical pendulum. Here, the pendulum means the set of the counterweight, arm, and thruster, and the other objects attached to the arm. Therefore, I includes the moment of inertial of thruster and counterweight.

For steady-state, the time variation in θ is ignorable, and then, the thrust can be evaluated using:

$$T = \frac{k}{l_t l_d} y_d \qquad (8.3)$$

Then, the sensitivity [m/N] is:

$$\frac{l_t l_d}{k} \qquad (8.4)$$

Hence, lowering spring constant increases sensitivity. However, this leads to a slower response, though a quick response is needed for the thruster that rapidly varies thrust. In other words, the sensitivity and response have trade-offs, and the analysis using the vibration theory can unveil the correlation between dynamic response and sensitivity. From here, the step response and frequency response will be shown to clarify the features.

Eqs. (8.1) and (8.2) are rewritten as the generalized form using the damping ratio ξ and natural angular frequency ω_n:

$$\ddot{y}_d + 2\xi\omega_n\dot{y}_d + \omega_n^2 y_d = \frac{l_d l_t}{I} T \qquad (8.5)$$

$$\omega_n = \sqrt{\frac{k}{I}} \qquad (8.6)$$

$$\xi = \frac{c}{2}\sqrt{\frac{I}{k}} \qquad (8.7)$$

When the step-like thrust (magnitude: T_0) is applied with initial conditions: $y_d(0) = 0$ and $\dot{y}_d(0) = 0$, the displacement is expressed as:

$$y_d = \begin{cases} \frac{l_d l_t T_0}{k}\left[1 - \exp(-\xi\omega_n t)\left\{\frac{\xi}{\sqrt{1-\xi^2}}\sin\sqrt{1-\xi^2}\,\omega_n t + \cos\sqrt{1-\xi^2}\,\omega_n t\right\}\right] & (0 \le \xi < 1) \\ \frac{l_d l_t T_0}{k}\left[1 - (1+\omega_n t)\exp(-\omega_n t)\right] & (\xi = 1) \\ \frac{l_d l_t T_0}{k}\left[1 - \exp(-\xi\omega_n t)\left\{\frac{\xi}{\sqrt{\xi^2-1}}\sinh\sqrt{\xi^2-1}\,\omega_n t + \cosh\sqrt{\xi^2-1}\,\omega_n t\right\}\right] & (\xi > 1) \end{cases}$$

$$(8.8)$$

The oscillations for $\xi < 1$, $\xi = 1$, and $\xi > 1$ are referred to as the underdamped, critically damped, and overdamped oscillations, respectively. Without any damper, i.e., $\xi = 0$, the oscillation never attenuates, as shown in Fig. 8.6. For $\xi = 0.2$, the amplitude still slowly decays, and the decaying time is decreased with increasing $\xi\omega_n$ owing to $\exp(-\xi\omega_n t)$. Accordingly, an increase in ξ slows down the rise time, whereas attenuating oscillation quickly.

For $\xi \ge 1$, the thrust stand yields no oscillation but extends rise time to deteriorate the response. Hence, under the condition where the oscillation needs to be entirely suppressed, the critical damping ($\xi = 1$) seems preferable since the rise

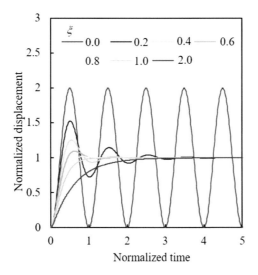

FIGURE 8.6

Time history of the displacement method. Time is normalized with $2\pi/\omega_n$ (the period at $\xi = 0$).

time is the lowest among the oscillation-less condition. (However, the critical damping yields a penalty in the phase delay. Optimum ξ requires further discussion, which will be mentioned later.) As shown here, the time histories presented in Fig. 8.6 provide some perspectives and are useful for discussing the overshoot and rise time but only provide slight information regarding the accuracy and upper-frequency limit of thrust measurement. Instead, the frequency responses offer many helpful insights.

Figs. 8.7 and 8.8 illustrate the frequency responses of the sensitivity and phase shift. The negative phase shift indicates the determined thrust delays compared with the actual value. The sensitivity and frequency are normalized with the sensitivity at 0 Hz (Eq. 8.4) and natural frequency, respectively; ideally, the sensitivity and phase should be 1 and $0°$. However, whereas the sensitivity and phase are kept at the ideal values for lower frequency, the sensitivity shows the discrepancy from the unity by increasing frequency, and the phase shift presents decreasing from zero, i.e., the thrust stand yields an increasing trend in phase delay. The sensitivity gives a peak for $\xi < 1$ and a monotonic decrease for $\xi \geq 1$, respectively. The upper measurable frequency can be determined using Fig. 8.7.

Table 8.1 summarizes the upper measurable frequency for each acceptable sensitivity error. The upper measurable frequency has a maximum at $\xi \approx 0.7$. For $\xi \leq 0.6$, the resonance increases overshoot to limit the upper measurable frequency, whereas, for $\xi > 0.7$, the damper suppresses the overshoot and yields resultant underestimation. Hence, ξ appears to have an optimum value neighboring 0.7 from the viewpoints of the measurable frequency. However, if the accuracy in both phase and sensitivity is necessary, ξ needs to be lowered because the phase delay reaches

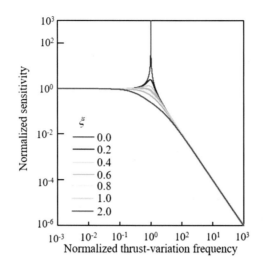

FIGURE 8.7

Dynamic response of the displacement method (sensitivity). Thrust-variation frequency is normalized with the natural frequency. Normalized sensitivity above 1 shows overestimation.

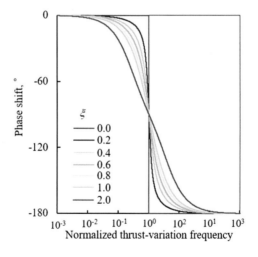

FIGURE 8.8

Dynamic response of the displacement method (phase). Thrust-variation frequency is normalized with the natural frequency. The negative sign of the phase shift indicates that the delay is induced.

$21.6°$ at $\xi = 0.7$. In contrast, at $\xi = 0.3$, errors in both phase delay and sensitivity become lower than 1% in the case that the thrust variation frequency is 10% of the natural frequency. Further decrease in ξ shortens the response time but increases

Table 8.1 Upper-frequency limit of measurable thrust variation from the view point of sensitivity. The upper limit frequencies are normalized with the natural frequency f_n.

Acceptable error in sensitivity							ξ					
	0	0.1	0.2	0.3	0.4	0.5	0.6	0.7	0.8	0.9	1.0
1%	0.097	0.098	0.101	0.107	0.121	0.146	0.195	0.263	0.059	0.040	0.032
3%	0.169	0.174	0.177	0.188	0.209	0.247	0.368	0.294	0.090	0.069	0.055
10%	0.301	0.305	0.316	0.337	0.378	0.473	0.769	0.396	0.185	0.127	0.101

the transient period to extend the time for gaining time-averaged thrust. From the discussion, the response and accuracy are the relations of the tradeoffs, and ξ should be determined in accordance with individual requirements for measurement.

When the thruster presents a time variation, please check the frequency response. Some papers used thrust stands with $\xi < 1$ and dealt with the overshoot as the thrust to result in overestimation. In other cases, the ignition delay was evaluated from the pendulum displacement and resultantly overestimated to deteriorate the attractiveness of the thruster. Hence, the frequency response must be cared about, primarily if the thrust presents time variation and rapid rise/fall in ignition or interruption.

Note that the load cell-based thrust measurement is theoretically identical to the displacement method, where the displacement sensor and springs are replaced with the strain gauges inside. The thruster mass lowers the natural frequency of the load cell-based thrust stand compared with that of the load-cell specification, even if the thruster is directly mounted on the load cell without any additional arm. A discussion in the frequency response is needed for the load-cell type either. Never blindly believe the measuremable frequency limit in the catalog. Analyze the loadcell-thruster systems using the vibration theory. Some manufacturers offer the stiffness of the load cells, and we can predict both the natural and upper measurable frequencies.

Interestingly, impulsive thrusts, which contain high-frequency components, can be evaluated, although the time history remains unknown. The impulsive thrust is determined under the condition the thrust production width is sufficiently smaller than the reciprocal of the natural frequency. The following section will address the impulsive thrust measurement.

8.1.2.2 Impulsive thrust

Impulsive thrust is also measurable with the displacement method, whereas the histories remain unknown [11,12]. In the case with the pendulum period $T = 2\pi/\omega_n$ is sufficiently longer than the thrust production duration, the pulsed thrust can be dealt as impulse. Under the impulse assumption, the initial velocity of the pendulum after the firing becomes:

$$\dot{\theta}(0+) = \frac{l_t I_t}{I} \tag{8.9}$$

For simplicity, assuming that $\theta(0) = 0$, the solution of Eqs. 8.1 and 8.2 is expressed as:

$$y_d = \begin{cases} \dfrac{l_t l_d I_t}{I \omega_n \sqrt{1 - \xi^2}} \exp(-\xi \omega_n t) \sin\left(\sqrt{1 - \xi^2}\, \omega_n t\right) & (\xi < 1) \\[2em] \dfrac{l_t l_d I_t}{I} t \exp(-\omega_n t) & (\xi = 1) \\[2em] \dfrac{l_t l_d I_t}{I \omega_n \sqrt{\xi^2 - 1}} \exp(-\xi \omega_n t) \sinh\left(\sqrt{\xi^2 - 1}\, \omega_n t\right) & (\xi > 1) \end{cases} \tag{8.10}$$

The equations show that the impulsive thrust I_t is proportional to the displacement amplitude, and hence, I_t is determined from the displacement amplitude. The underdamped condition is usually applied to the thrust stand to enlarge the sensitivity. For further enhancement of sensitivity, the spring constant k, momentum of inertia I, and damping coefficient c need to be reduced as possible since the denominator has $I\omega_n = \sqrt{KI}$ and $\sqrt{1-\xi^2}$. However, lowering c, i.e., ξ, extends the period for attenuating the pendulum motion to prolong wait time to the next firing. Then, it would be helpful to adjust the damping coefficient in order to reduce the attenuating period without deteriorating sensitivity: weakening and strengthening the damping coefficient during and after thrust measurement, respectively.

8.1.3 The null-balance method

The null-balance method is applied to the thrust stand [13−16]. In general, this is the same as the balance scale in principles; position and number of known-mass weights are adjusted such that the balance scale becomes horizontal, and the target-object mass is evaluated using the position and total mass of the weights. The method covers various types of sensors: accelerometers and force sensors. In the thrust measurement, the weights and the total mass are equivalent to the actuator and the controller output (actuator-driving current or voltage). The section addresses the theory, practice, and case study for evaluating frequency response.

8.1.3.1 Theory
The angular equation of motion for the null-balance method using the pendulum model shown in Fig. 8.9 is expressed as:

$$I\ddot{\theta} + c\dot{\theta} + k\theta = l_t T + l_a F(\theta, u) \tag{8.11}$$

$$y_d = l_d \theta \tag{8.12}$$

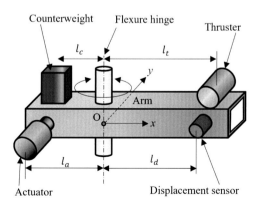

FIGURE 8.9

A dynamical model of the null-balance method.

where F and u are the actuator force and the controller output, respectively. Let us assume that the electromagnetic actuator is used since the electromagnetic actuator is comparatively simple and robust and produces a wide range of force. If the nulling control constraints the pendulum position to the vicinity of the target position, the force is proportional to the driving current so that $F = F_0 i$. Then:

$$I\ddot{\theta} + c\dot{\theta} + k\theta = l_t T + l_a F_0 i \tag{8.13}$$

Hence, when the controller zeros the displacement, the thrust is evaluated using:

$$T = -\frac{F_0 l_a}{l_t} i \tag{8.14}$$

Accordingly, the sensitivity S [A/N] is:

$$S = -\frac{l_t}{F_0 l_a} \tag{8.15}$$

The method can provide the following features.

- The actuators can play the roles of springs and dampers, and hence, mechanical dampers and springs, which sometimes yield hysteresis, are not necessary.
- The frequency response, sensitivity, and the zero-thrust position of the pendulum are flexibly adjustable by the controller tuning, though the displacement method occasionally requires the replacement of the springs, torsional hinges, and mechanical dampers.
- Even if a nonlinear displacement sensor is used, the thrust stand maintains the linearity between the thrust and controller output (actuator-driving current). Since some nonlinear displacement sensors have finer resolution than the linear displacement sensors despite the lower price, the application of the null-balance method readily enhances the accuracy and resolution.

8.1.3.2 Controller design
The classical proportional-differential-integral (PID) control is usually used to null the displacement. When voltage-based driver is applied, the control voltage V is determined from

$$V = k_P \theta + k_D \dot{\theta} + k_I \int \theta dt \tag{8.16}$$

From Kirchhoff's law,

$$V = L_a \frac{di}{dt} + R_a i \tag{8.17}$$

where the target position is set at $x = 0$, and L_a and R_a are the equivalent inductance and resistance of the electromagnetic actuator. Then, the thrust is determined from V. When the amplifier regulates current, the current is expressed using

$$i = k_P \theta + k_D \dot{\theta} + k_I \int \theta dt \tag{8.18}$$

The voltage-based control seems to yield a slower dynamic response than the current-based control because the inductive-resistance circuit (LR circuit) produces the first-order delay. However, the discrepancy is negligible owing to the shorter time constant of the actuators. For electromagnetic actuators, the time constant is L_s/R_s for the voltage-based control, and L_s and R_s are on the order of 10 mH and a few Ω, respectively so that the time constant is on the order of ms. On the other hand, the thrust stand pendulums have a natural frequency neighboring 10 Hz under the null-balance control. Hence, the actuator circuit responds quicker so that the first-order delay of the LR circuit is ignorable.

In design, the coefficients k_P, k_D, and k_I need to be determined to stabilize the nulling control. Improper coefficients generally destabilize control systems: the pendulum keeps oscillating or the displacement diverges. For stability, gain and phase margins are occasionally used as the practical criteria. The Bode plot of the loop transfer function provides these margins and gives the perspectives of the control-system dynamics. Eqs. (8.13)–(8.18) are so simple that the Bode plot can be drawn by hand calculation. However, the calculation is elaborating, and hence, the use of simulation software such as MATLAB and Scilab may be the practical option.

The use of the D element, which can stabilize the control systems, requires low pass filters. The sensor signals inherently contain noise, and then, the differentiation of the signals always amplifies the high-frequency noise so that the null-displacement control fails. Hence, low-pass filters are mandatory for stable control. The lower-order analog filters are readily fabricated using only a few operational amplifiers, capacitors, and resistors. However, higher performance filters that have a steep roll-off complicate the circuit design.

Instead of the analog filters, the computers can filter the sensor signal. The finite impulse response (FIR) filter would be a promising option from the viewpoint of stability, design simplicity, and calculation costs among various filters. The N-th order filter is presented as

$$y[n] = \sum_{i=0}^{N-1} b_i x[n-i] \qquad (8.19)$$

where $x[n]$ is the input (the newest signal), $x[i]$ ($i < n$) is the past sampled signal, and $y[n]$ is the output. The filter can become low-pass, band-pass, and high-pass filters by selecting b_i. The filter coefficient b_i ($0 \le i < N$) can be determined using various methods, and MATLAB and Scilab offer designated functions to calculate the filter coefficients.

The PID controller with filters can be implemented using analog circuits with operational amplifiers. However, the circuits would become complicated when sophisticated signal processing, such as higher order low-pass filtering, is required. In such cases, computers can be an appropriate option. Complex calculations, including the nonlinearity compensation and high-order low-pass filtering, can be implemented using software or field-programmable gate arrays (FPGAs). Moreover, the use of computers and FPGAs allows flexible tuning of the parameters for the PID control and filter.

When laptop/desktop computers are applied, care about the jitter. Windows, the widely used operating system (OS), is not a real-time OS and does not ensure the accuracy of periodic processing. In other words, in the critical routine control, the Windows-based computers sometimes produce unacceptable jitter, which is generally non-predictable. Hence, the real-time OS should be used for higher frequency response and accuracy. Linux-kernel variants such as RTAI, Xenomai, and Preempt-RT are promising options and the open-source software. For a 1 ms periodic routine, the jitter was below 40 μs. The commercial real-time OSs are said to produce lower jitter, but the additional expense would be needed.

The cutting-edge microprocessors for the embedded systems would be applicable to the controller. Recent technology development allows the embedded CPUs to have floating-point units (FPUs), which quickly calculate the floating-point values. Devices referred to as a system on chips (SoCs) have not only CPUs and FPUs but various interfaces (analog interface and communication ports). The SoCs are manufactured to be used in various commodities so that the price is reasonable, and the performance and function are sufficient enough to determine the controller outputs and manipulate actuator driving currents. Hence, these devices are a promising option.

As described in the section, the design and implementation of the controller require some knowledge regarding control theory and electric circuits, but there are proper design-assistant tools. Hence, the barrier to the controller design is readily broken with the assistance of such tools and open-source software.

8.1.3.3 Dynamic response

The sensitivity of the null-balance method yields a similar frequency response to that of the displacement method. Though the analysis using generalized parameters ξ and ω_n is ideal for gaining insight into the system dynamics, the section uses a case study owing to the difficulty in the generalization. Fig. 8.9 and Table 8.2 show a model and configuration, respectively. The voice coil motor (VCM) specification is the same as that of VCMs that the authors prototyped for the magnetic-levitation thrust stand that measures the six-component thrust vector [17]. The counterweight mass was adjusted such that the C.G. position of the pendulum (thruster, counterweight, and arm) is located at the hinge. As will be mentioned later, the adjustment of the C.G. position can suppress the error due to the external vibration. With the model, numerical simulation was conducted using Scilab.

Fig. 8.10 illustrates the pendulum displacement amplitude for thrust variation 1 N in amplitude. The maximum displacement amplitude is approximately 1.1×10^{-3} m/N at 7.1 Hz and originates from the resonance. In the PID control, k_P and k_D play the spring and damper roles, respectively, and hence, the null-balance method yields a similar frequency response to those of the displacement method to generate the amplitude peak. On the other hand, for the lower frequency

Table 8.2 Configuration of the thrust stand model.

Arm	Geometry	Square pipe
	Size, m^3	0.05 × 0.5 × 0.75 (5 mm thickness)
	Material	Duralumin
VCM	Resistance, Ω	2.5
	Inductance, mH	10
	Position x_a, m	0.5
	Force to current ratio F_0, N/A	1.0
Thruster	Mass, kg	0.5
	Position x_t, m	0.5
Controller	Types	Digital control
	Sampling rate, Hz	1000
	k_P	1000
	k_D	10
	k_I	10
Counterweight	Position x_c, m	−0.25

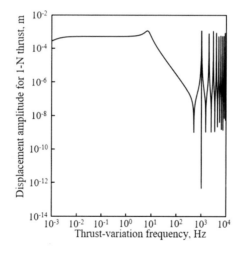

FIGURE 8.10

Frequency response of the displacement for 1-N thrust variation.

region, displacement amplitude is deceased with decreasing thrust variation frequency. The reduction is attributable to the I element, which contributes to the attenuation in residual error in the displacement, especially in a lower frequency range.

Below the resonant frequency, the displacement ranges from 2.67×10^{-4} to 1.1×10^{-3} m/N. Since the reasonable-price laser displacement sensors have resolutions of approximately 10 μm, the thrust stand can be implemented using available laser displacement sensors. Above 500 Hz, a bottom and peak appear at 500 and 1000 Hz, respectively, and emerge every 1000 Hz. The variations originate from the digital control, and the 1000-Hz interval comes from the sampling rate. As shown later, the frequencies at the peaks and bottoms are higher than the measurable frequency, and then, no influence is provided to thrust determination.

Here, the accuracy will be discussed using the phase shift and normalized sensitivity (the amplitude ratio of the evaluated thrust to the actual thrust), the ideal values of which are 0° and unity. Figs. 8.11 and 8.12 present the frequency responses of normalized sensitivity and phase shift, respectively, and measurable frequency ranges are summarized in Table 8.3. As in the displacement method with $\xi < 1$,

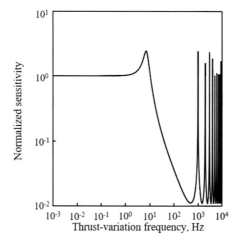

FIGURE 8.11

Normalized sensitivity.

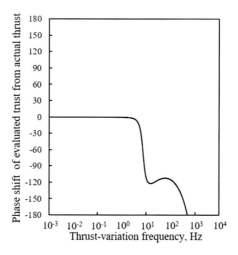

FIGURE 8.12

Phase shift of evaluated thrust from the actual thrust.

Table 8.3 Measurable upper frequency of the null-balance method model (Fig. 8.9 and Table 8.2). Sensitivity is normalized with that at 0 Hz. Normalized sensitivity above 1 shows overestimation.

Acceptable error in sensitivity	Measurable upper frequency, Hz	Corresponding phase shift, °
1%	0.759	−0.165
3%	1.29	−0.374
10%	2.29	−1.25

the frequency response of the sensitivity has peaks and bottoms. Below a thrust variation frequency of 0.759 Hz, the thrust stand showed a sensitivity error and phase delay of below 1% and 0.165°. The sensitivity error and phase delay are increased with increasing thrust-variation frequency. If acceptable errors in sensitivity are 3% and 10%, the upper measurable frequencies are 1.29 and 2.29 Hz with the corresponding phase delays of 0.374 and 1.25°. Hence, for practical use, 0.759−2.29 Hz would be the upper measurable frequency. Afterward, the sensitivity error continuously increased with thrust variation frequency and has the first peak at 7.1 Hz, and phase delay keeps enlarging. Then, the sensitivity had peaks and bottoms every 1000 Hz above 500 Hz. Such peaks have no influence on thrust determination because the frequency is much higher than the measurable frequency range. On the other hand, the phase delay is monotonically increased with increasing thrust variation frequency.

From these results, the null-balance method can evaluate thrust and have an upper measurable frequency. The controller tuning (adjusting PID parameters) can extend the frequency limit, and the augmentation in P and D enhances the response. However, this increases the resonant frequency originating from the control, and hence, requires a higher controller sampling rate. The pendulum displacement is also reduced, and thus, stable control requires a finer-resolution displacement sensor. Moreover, the thrust stand parts would have to have more substantial stiffness because the controller roughly adjusts the actuator to induce the elastic vibration possibly. Hence, practically, the upper measurable frequency would be restricted to tens of Hz. The application of acceleration measurement to the null balance is attempted to extend the frequency range [18] and will be introduced in Section 8.1.7.

8.1.4 Thrust target

The thrust target has been used when the thrust-to-weight ratios are so small that the thrust stand cannot evaluate thrust, or the tubes and wires are so inflexible that they present hysteresis or reduce the pendulum-arm displacement to the undetectable range [19−21]. Fig. 8.13 shows a schematic of a typical thrust target. In the method, the thrust target, which is fixed with a flexural hinge, is collided with the thruster exhaust jet to start oscillating. From now on, the feasibility is discussed theoretically.

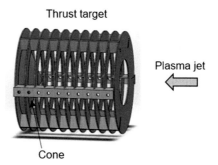

FIGURE 8.13

A thrust target.

The thrust target is modeled, as illustrated in Fig. 8.14. The conservative laws of mass and momentum are expressed as:

$$\frac{\partial}{\partial t}\int \rho\vec{u}\,dV + \int \rho\vec{u}\left(\vec{u}\cdot\vec{n}\right)dA = \sum\vec{F} \tag{8.20}$$

$$\frac{\partial}{\partial t}\int \rho\,dV + \int \rho\vec{u}\cdot\vec{n}\,dA = 0 \tag{8.21}$$

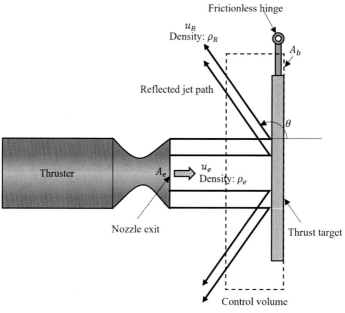

FIGURE 8.14

A model of the thrust target.

where V is the volume of the control volume, A is the control surface, $\vec{n}$ is the unit normal vector of the small control surface dA, $\vec{u}$ is the jet velocity, and ρ is the mass density. Assuming that the viscosity is ignorable on the control surface, the thrust target provides the force F'_{tp} to the jets so that Eq. (8.20) is converted into:

$$\frac{\partial}{\partial t}\int \rho \vec{u}\, dV + \int \rho \vec{u}\left(\vec{u} \cdot \vec{n}\right) dA = \vec{F'}_{tp} - \int \rho \vec{n}\, dA \tag{8.22}$$

where p is the pressure. For the steady thrust thruster, the first term of the left-hand side is eliminated, and for simplicity, the system is assumed to be axially symmetric. Then, the z-axis component is:

$$-\rho_R u_R^2 A_R \cos\theta - \rho u_e^2 A_e = F'_{tp,z} + p_e A_e + p_a(A_b - A_e) - p_a A_b \tag{8.23}$$

$$-\rho_R u_R A_R \cos\theta - \rho_e u_e A_e = 0 \tag{8.24}$$

According to the law of the action and reaction, $F_{tp,z} = -F_{tp,z}'$. Then:

$$F_{tp,z} = \rho u_e^2 A_e + A_e(p_e - p_a) - \rho_e u_e A_e u_R \cos\theta \tag{8.25}$$

Here, the thrust of the rocket is the sum of the momentum thrust and pressure thrust:

$$T = \rho u_e^2 A_e + A_e(p_e - p_a) \tag{8.26}$$

Therefore, we gain:

$$F_{tp,z} = T - \rho_e u_e u_R A_e \cos\theta \tag{8.27}$$

Hence, the impulse gained by the thrust target equals the impulsive thrust in the case with $\theta = 90°$ (the jet is expelled in the radius direction), or $u_R = 0$ (decelerating the jet to zero). Because of the difficulty in realizing $u_R = 0$, the practical approach is $\theta = 90°$. The thrust target in Fig. 8.14 has a cone and rings to let the exhaust turn perpendicularly [19]. If the jet is rebounded to go back to the thruster, i.e., $\theta > 90°$, the thrust would be overestimated. The worst case is that the jet is bounced off the target through the perfectly elastic collision and maintained the velocity so that $\theta = 180°$ and $u_R = u_e$. In this case, the evaluated impulse doubles the actual value.

Moreover, the high-speed exhaust jet impinges on the thrust target, some of which may evaporate. The evaporated target would expand to exert aerodynamic forces and cause errors. The complex process in the jet-target impingement and target evaporation perplex the error analysis. Accordingly, the thrust stand should be used as possible, and the use of the thrust target should be used as the second-best measure.

8.1.5 Elements

8.1.5.1 Materials
Discussion in Sections 8.1.2 and 8.1.3 assumed that all the materials, including the pendulum arms, are the rigid body and ignores elasticity. However, the elastic vibration can be induced in any real parts. For instance, the arms with thrusters and

counterweights are equivalent to the cantilever supporting weights, which readily induces transverse and torsional vibrations. Wires, with which pendulum arms are sometimes suspended, readily induce longitudinal vibrations as well as transverse and torsional vibrations and have lower natural frequencies than beams owing to the lower diameter.

Such vibrations are sometimes harmful to thrust measurement. If the natural frequency neighbors thrust-variation frequency or the controller sampling rate, the induced elastic vibration is superimposed into the thrust stand outputs to become the mechanical noise. Moreover, the elastic vibration may destabilize the nulling control in the null balance method. Note that the step and delta functions contain wide-frequency components. The thrust is sometimes pulsed with ms width and, even for steady firing, is suddenly increased and decreased in thrust at ignition and termination. Regarding thrust stands, the digital control repeatedly adjusts the solenoids to produce a train of pulsed forces whose width is the same as the reciprocal of the sampling rate. Hence, both the thruster and digital nulling control provide step-like and pulsed forces and could excite unexpected elastic vibrations.

From the discussion, modal analysis or theoretical calculation of the natural frequency of the vibration is sometimes necessitated in thrust stand design, and some research groups conducted the FEM simulation [22,23]. Fig. 8.15 shows an example modal analysis result, which was conducted using finite element method (FEM) software bundled with 3D computer-assisted design (CAD) software. The thruster is modeled as 1 kg rigid-body tubes; the frames are hollow duralumin beams; two flexural hinges are installed. The first natural frequency is as low as 70 Hz, though a supporting plate is installed to the frame. In general, the natural frequency of elastic vibration is lowered to the order of 10 Hz. Hence, the shape and elastic modulus need to be considered because it affects the natural frequency. In Ref. [22], the FEM was conducted to ensure that the flexure plate hinge produces a sufficiently low spring constant and that the frames are not more flexible than the flexure plate hinge.

FIGURE 8.15

A result of FEM vibration analysis for a thrust stand.

On the other hand, weight is another essential factor but has a trade-off relation to the natural frequency. Hence, for both lightweight and stiffness, modal analysis and weight assessment are necessary. Some open-source CAEs such as Salome and Elmer or 3D-CAD-bundled CAEs provide helpful insights in design. Even without such FEM tools, the weight can be precisely calculated, and approximate natural frequency is determined, assuming that the parts are the elastic beams with point masses.

8.1.5.2 Hinges

Hinges are usually used in the thrust stand. However, some hinges yield hysteresis and friction, and hence, should be carefully selected because friction and hysteresis deteriorate the accuracy. An option is a flexural hinge (referred to as flexural bearing, flexural pivots, or Flexi hinge). As shown in Fig. 8.16, the hinge is comprised of two or three tubes and planar springs, and the tubes are connected through the planar spring. The design allows the production of restoring force without hysteresis or nonlinearity. The hinge is tolerant of the bending moment induced by the pendulum and thruster weight and useable in the vacuum environment. However, some of them used ferromagnetic materials and hence yielded measurement errors if the thruster emits magnetic fields.

Note that such a flexure hinge presents transverse flexibility in nature. Fig. 8.17 shows a typical deflection analyzed using the static force FEM. This means that the thrust stand arm may induce the resonance in the transverse direction, which disturbed thrust measurement [23]. The transverse oscillation may deteriorate impulsive thrust evaluation. Moreover, the flexure hinge changes the spring constant owing to the torque perpendicular to the rotational axis. When the C.G. position is far from the rotational axis, the torque is induced and changes the spring constant. As long as the torque is within the rated value, no damage is provided to the hinge but changes the natural frequency of the pendulum arm.

Knife edges, which present negligible frictional forces, are a simple and effective hinge and applicable to the vertical pendulum type thrust stands. Though the hinge has no flexible element that produces restoring torque, the restoring force is produced by gravity. Sharp-edged triangles or knife-like plates are used because of their simplicity in machining. Moreover, the structure of the pendulum arm would be

(a)Flexural hinge   (b) Inner structure

FIGURE 8.16

A flex-ural hinge.

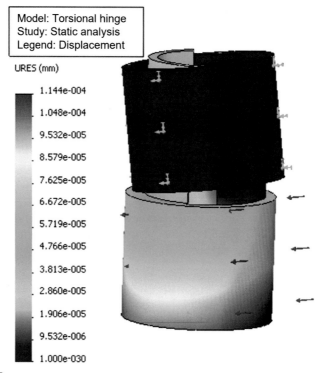

Model: Torsional hinge
Study: Static analysis
Legend: Displacement

URES (mm)

1.144e-004
1.048e-004
9.532e-005
8.579e-005
7.625e-005
6.672e-005
5.719e-005
4.766e-005
3.813e-005
2.860e-005
1.906e-005
9.532e-006
1.000e-030

FIGURE 8.17

FEM result of flexural pivot [23].

uncomplicated compared with that using the flexural hinges. However, the rotational axis can be shifted by the thrust or the forces produced during the experimental-apparatus setting, and further, the thrust stand has some difficulty in setting the counterweight compared with the flexure hinge. Therefore, the selection of hinges is dependent on the thruster weight and required sensitivity and accuracy (the necessity of counterweight).

8.1.5.3 Dampers

The dampers are necessary to suppress the pendulum oscillation induced by the sudden change in thrust. Without dampers, the thrust stand arm keeps oscillating because there is no drag force in the vacuum or frictional force in the well-designed hinges. Hence, various dampers are applied to thrust measurement.

Oil dampers can provide sufficiently large damping forces, which allow the thrust stand to present even the underdamped oscillations. The damper usually has a piston and cylinder, and the damping coefficient is theoretically predictable. The damping coefficient is dependent on the gap between the piston and cylinder, piston length, and oil viscosity. When thrust is measured in a vacuum chamber, the vapor pressure of oil should be lower than the vacuum chamber pressure, or the damper needs sealing, which could induce hysteresis and friction. Moreover, strengthening damping tends to yield hysteresis and nonlinearity. Hence, the damper is a reliable option, but care is needed for the stronger damper and the use in vacuum.

Passive magnetic dampers are also used in the thrust measurement (Fig. 8.18). The damper uses the eddy current, which is induced in a moving conductor under a static magnetic field, to exert damping force on the conductor by the interaction of the eddy current with the magnetic field. The damper creates no friction owing to the noncontact nature. The conductor is usually fixed on the thrust stand pendulum, and a magnetic circuit is placed out of the pendulum because the latter is more massive than the former. Copper is a promising conductor owing to the price and relatively high conductivity since an increase in electric conductivity augments drag forces. Neodymium magnets are also used due to strong magnetomotive force. The magnetic damper requires no electric power or control system and produces no frictional force. Whereas the precise design could necessitate some complicated numerical simulation, some simplified theoretical equations are proposed.

FIGURE 8.18

An example of an eddy current damper.

Active dampers, which use the actuator and displacement or velocity sensor, can provide damping forces. When the displacement sensor is used, the sensor output signals are differentiated with regard to time to gain velocity, which needs to be filtered since the differentiation of the measured value amplifies the noise. The actuator is driven with the current or voltage that is proportional to the velocity. The damper shows no hysteresis but yields divergent oscillation when the damping coefficient or filtering is inappropriate. In the calculation, both analog circuits and computers are applicable to the differentiation and filtering. The features of actuators will be discussed in Section 8.1.5.5.

8.1.5.4 Displacement sensors

The pendulum displacement is mandatory for both the displacement and null-balance methods, and the sensor is required to have the features: temperature stability, accuracy, fine spatial resolution, frictionless, and low hysteresis. Linearity seems not always necessary, but nonlinearity sometimes requires careful treatment. From the viewpoints, laser displacement sensors, linear variable differential transformers (LVDTs), and eddy current sensors have been used in the thrust measurement.

The laser displacement sensor has been applied owing to appropriate features. There are various types: triangulation, optical knife-edge, interferometry, and time of flight types. From the viewpoint of price, usability, and availability, the triangulation type is widely used. In the sensor, the semiconductor laser emits laser beams, and a linear-array sensor like a linear CCD receives the reflected laser beam. The distance is determined from the position of the linear-array sensor where the reflected laser beam arrives. The spatial resolution ranges from nm to mm with a measurable range of millimeter to meter, and the sampling rate can be on the order of ms. The noncontact nature provides a flexible selection of the fixing point, and the opposing object is freely selected as long as the object reflects the laser. Hence, the laser displacement sensor is an appropriate option. The light-emitting diode (LED) type is also used because the sensor has the same features as those of laser displacement sensors with lower prices. However, the LED type is worse in resolution than the laser type.

The LVDT has preferable characteristics. The sensor, which is a transformer that has a movable ferromagnetic core without electric parts such as amplifiers, as shown in Fig. 8.19, is robust and applicable to cold and vacuum environments. The primary coil is driven with an alternate current of $1-20$ kHz in frequency, and the amplitude of the secondary voltage is proportional to the cylindrical ferromagnetic core. Some LVDT signal conditioners smooth the alternate current from the secondary to output the voltage or current that is proportional to the displacement. The μm-order spatial resolution is not uncommon. In the selection, the gap between the coil and ferromagnetic core needs to be considered because the core is inserted into the coil.

Eddy current sensors detect the displacement using eddy currents. The sensor induces a high-frequency magnetic field using the coil and then produces the eddy current in an opposing metal. The eddy current changes the impedance of the coil, which depends on the gap between the sensor and the metal. The sensor output is

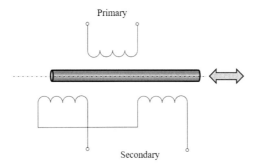

FIGURE 8.19

A linear variable differential transformer.

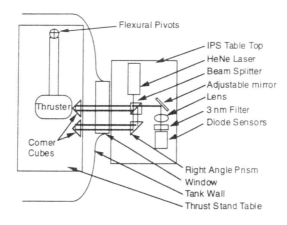

FIGURE 8.20

A laser interferometry displacement sensor [24] that is dependent on the displacement. The sensor used in Ref. [8] had a spatial resolution of 25 nm.

conditioned with the signal conditioner so that the voltage or current signal is output. The sampling rate is as high as 40,000 Hz, and spatial resolution can be on the order of nm. Note that the opposing objects should be metal.

Laser interferometry was applied to impulsive thrust measurement [24]. The laser interferometry proximeter system, illustrated in Fig. 8.20, yielded a spatial resolution of 10 nm, which is finer than other sensing devices. Interference, induced by the difference in the optical path, changed joined-beam intensity, which is dependent on the distances between the thruster and optics (prism and beam splitter) so that the displacement was evaluable from the intensity. The resolution was sufficient for the performance evaluation for pulsed plasma thrusters, which produced impulsive thrust on the order of 10 μNs.

Capacitance displacement sensors were used because of linearity, quick response, and fine resolution (Fig. 8.21). For the capacitors, the capacitance is

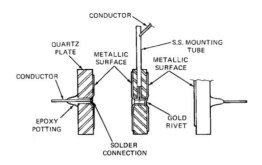

FIGURE 8.21

A capacitance-type displacement sensor [25].

reciprocally proportional to the distance. In some sensors, a sinusoidally varying current was provided, and the capacitor presents alternating voltage with an amplitude.

Aside from these sensors, nonlinear sensors can be an option owing to lower price and nm-order resolution. A nonlinear sensor, which forms inductance and capacitance oscillating circuit an opposing metal, evaluates distance from the oscillating frequency. Whereas such a sensor seems inconvenient in the displacement method, the null balance method can maintain linearity between thrust and output despite nonlinearity so that the use of the sensor enhances accuracy and frequency response.

8.1.5.5 Electric actuators

The use of electric actuators can eliminate some mechanical parts: springs, dampers, and the adjusters that adjust the pendulum-equilibrium position, and hence, are applicable to both the displacement and null-balance methods. The actuator can produce reference thrusts in calibration. The actuator should have a quick response, frictionless due to its noncontact nature and sufficient force.

The VCM, the electromagnetic actuator that is comprised of a yoke (stator) and voice coil, as shown in Fig. 8.22, is a promising option since the force is proportional to the driving current and independent of the gap between the voice coil and yoke. The response time is fast among the actuators because the voice coils are relatively lightweight. If reducing the thrust-to-current ratio is accepted, the VCM can have longer gaps between the stator and voice coil, which is beneficial for assembling the thrust stand. Moreover, VCMs can be readily designed using the FEM. Open-source or free FEM simulation software packages are available: Finite Element Method Magnetics (FEMM) and Quick field. Such software predicts not only magnetic field distribution but also force-to-current ratio, inductance, and resistance.

Solenoid type actuators can yield stronger forces than VCMs. The actuators have various structures: the permanent magnets are inserted into solenoids, whereas the ferromagnetic cores are inserted into solenoids. However, the force is a nonlinear function of the gap between the magnet and the solenoid, whereas being linear

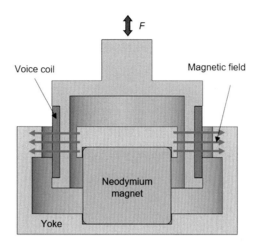

FIGURE 8.22

A schematic of a voice coil motor [17].

with the driving current. The nonlinear dependence could destabilize the control in the null-balance method and distort the time history of the pendulum displacement in the displacement method. As in VCMs, the FEM software can anticipate the dependence of the force on the gap and current.

Aside from these actuators, electrostatic actuators are also used if the thrust is sufficiently small. The actuator uses the phenomena where electrodes attract each other in a capacitor. The structure and principle are straightforward, and the force is adjusted with the voltage. The electrostatic calibrator will be introduced in Section 8.1.6.

8.1.5.6 Counterweight

The counterweight is usually placed such that the C.G. of the pendulum is located on the hinge (rotational axis). The management of C.G. can suppress the mechanical noise induced by the external vibration transmitted by the vacuum chambers or tables where the thrust stand is settled. The chapter discusses the effect of the counterweights. When the thrust stand was vibrated at the acceleration (a_x, a_y) through the vacuum chamber or table vibration, the equation of the angular motion that includes the inertial force produced is expressed in the equation using the C.G. position (x_G, y_G) with the corresponding angle θ_0 and length between C.G and hinge:

$$x_{G0} = l_G \cos \theta_0 \tag{8.28}$$

$$y_{G0} = l_G \sin \theta_0 \tag{8.29}$$

$$I\ddot{\theta} + c\dot{\theta} + k\theta = l_t T + l_a F_0 i + m\{l_G \sin(\theta + \theta_0)a_x - l_G \cos(\theta + \theta_0)a_y\} \tag{8.30}$$

where m is the total pendulum mass, the sum of masses of a pendulum arm, thruster, and counterweight. The equations show that whereas the inertial force negatively affects thrust measurement, diminishing the offset l_G can eliminate interaction. In other words, the counterweights can isolate the rotational motion from the translational motion. Hence, the counterweights are located on the other end of the thrust stand pendulum, as illustrated in Fig. 8.5.

If the thrust is tiny, the counterweight position needs to be carefully adjusted since a slight error in the C.G. position leads to the nonnegligible measurement noise. One of the major problems in adjusting C.G. position is that there is no definitive and straightforward method to evaluate the offset of C.G. position. Providing external vibration at the natural frequency of the thrust stand pendulum $\frac{1}{2\pi}\sqrt{k/m}$ would unveil the C.G. position offset because inducing the pendulum displacement when the C.G is not on the rotational axis. Then, the counterweight mass and position are adjusted such that no rotational motion is generated. If the inaccuracy is acceptable, a simple way is to tilt the thrust stand; if there is an offset of the C.G. position, the pendulum arm drifts from the equilibrium position. Then, adjust counterweight mass and position such that tilting induces no pendulum-arm displacement.

8.1.6 Calibration

The calibration is indispensable to the thrust measurement. Although the sensitivity is theoretically predictable, the shift in C.G. position and flexible elements such as tubes and wires exert external forces so that the sensitivity is different from the prediction. Therefore, to evaluate the sensitivity, the thrust stand should be calibrated under the same conditions as those of firing tests. Reference thrusts that certified instruments determine should be applied and varied up to the maximum thrust of the test thruster. There are various methods for calibration. This section deals with the calibration devices and methods (Table 8.4).

Table 8.4 Calibration methods.

Force source	Applicable thrust	Evaluating reference force
Weight	Steady	Mass balance
Pendulum-type impulse hammer	Impulsive	Load cell, velocimeter, mechanical energy conservation
Electromagnetic	Steady/impulsive time-varying	Mass balance, load cell
Electrostatic	Steady/impulsive time-varying	Mass balance, load cell
Gas jet	Steady/impulsive	Nozzle theory

8.1.6.1 Steady thrust

Weights and pulleys have been frequently used for calibration. Fig. 8.23 illustrates a weight-based calibrator where the mass-evaluated weights (reference weights) are tied with a string, which is connected to the pendulum arm through a pulley. A movable stage lifts the reference weights, and the number of the weights that exerted the reference force is adjustable by the movable-stage height. The method is simple in theory and practice and readily implementable. A series of reference weights and an elevator are used to vary the reference thrust.

Electromagnetic actuators have also been used to provide the reference thrusts and are found in Ref. [8]. Figs. 8.24 and 8.25 shows electromagnetic calibrator w/and w/o load cell, respectively [17,27]. Different from the weight type, the calibrator can provide both steady and impulsive thrusts and hence, applicable to steady/impulsive thrust measurement. Moreover, applying time-varying reference thrust enables the evaluation of frequency response. Such electromagnetic actuators

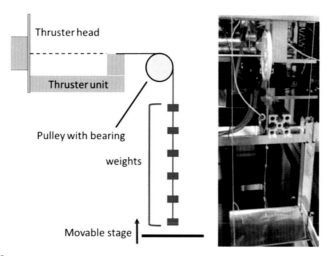

FIGURE 8.23

An example of the pulley and weight calibrator [26].

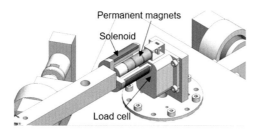

FIGURE 8.24

A permanent-magnet-solenoid calibrator [17].

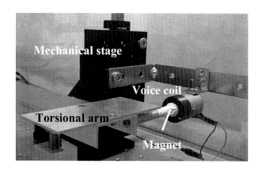

FIGURE 8.25

A voice-coil-motor calibrator [27].

have no contact between moving and stationary parts, and their noncontact nature eliminates friction and hysteresis. The force is readily adjustable with driving current and turns of the solenoid, and the calibrator can cover a wide range of thrust levels. The structure and principle are straightforward, and hence, the design is relatively easy. The magnetic analysis software using FEM, such as FEMM and Elmer, can predict the performance to assists the design.

Some calibrators, such as Fig. 8.24, used permanent magnets and solenoids and show the dependence on the gap between the calibrator and the thrust stand. The use of the load cell eliminated the care about the gap. The author designed the calibrator (Fig. 8.24) to evaluate the six-component thrust vector measurement [17]. The calibrator exerted not only the steady reference thrust up to 1 N but also sinusoidally varying thrust so that the calibration showed that the prototype evaluated six-component thrust vector with sufficient linearity and measured thrust variation up to 3 Hz.

VCMs are also applied to the calibrator (Fig. 8.25), and the thrust is independent of the gap between the yoke and voice coil and linear with the driving current. Hence, if the pendulum moves owing to the reference force, the calibrator can keep exerting the same force owing to the force independence on the gap. Lam et al. designed and tested the VCM-based calibrator and showed that the calibrator provided reference thrust in the range of 30−23,000 μN with a maximum uncertainty of 18.47% [27].

Note that from the viewpoint of traceability, the calibrator itself must be calibrated using some certified instruments such as mass balance, though the force-to-current ratio is theoretically predictable. This is partly because the solenoid requires neat wounding; otherwise, the force-to-current ratio is different from the theoretical value.

Electrostatic actuators are also used in the calibrator [12,28,29]. The actuator forms capacitance and yields force by applying voltage to the electrodes. The reference thrust is adjustable with a driving voltage because the electrostatic force is dependent on the square of the applied voltage. Then, time history is flexibly adjustable by varying driving voltage as in the electromagnetic actuator.

There are various electrode configurations, and one of the simplest configurations is the parallel plate. J. Lun and C. Law designed a thrust stand to evaluate μN steady and μNs impulsive thrusts [28]. The calibrator used the electrostatic disks to offer the reference. Calibration using the calibrator shows that the stand can measure a steady thrust of 27−600 μNm, which corresponds to 0.27-600 μNs repetitive impulsive thrusts for 1−100 Hz in firing frequency. GameroCastaño prototyped μN thrust stand with better than 0.03 μN resolution for μN thrusters, calibrated using parallel plate electrodes, and proved that the stand determined thrust ranging from 2 to 13 μN with a noise level of below 0.1 μN/$\sqrt{\text{Hz}}$ [12]. The calibrator produced sinusoidal reference thrusts to show that the thrust stand had a resonant frequency of 0.25 Hz with a frequency response similar to that with 0.4-Hz cut-off frequency.

Some groups prototyped fin-type electrodes. Cheah et al. prototyped the fin-type electrostatic calibrator presented in Fig. 8.26, which provided steady reference thrusts of 30−3400 μN by the application of approx. 150−900 V, with a deviation error of less than 2% [29]. The paper clarified the dependence of the electrostatic force on the number of fins and the correlation between the force and the interelectrode gap.

Gas jet is also used as the calibrator to evaluate nN class thrust [30]. Figs. 8.27 and 8.28 present schematics of the gas jet calibrator and the orifice for injecting the gas jet. The calibration used the direct Monte Carlo simulation (DSMC) and theoretical evaluation using underexpanded free jets with an assumption of the free molecule flow. Theoretical calculation with free molecule flow assumption yielded the same results as those with DSMC and hence, can be the reference thrust. Moreover, the tests showed that the LVDT output was proportional to the stagnation pressure of the gas jet. The thrust stand measured a thrust of 86.2 nN with an estimated accuracy of 11% and 712−1000 nN with an estimated accuracy of 2%. The gas jet calibration was compared with the electrostatic calibration [31]. The impulse discrepancy between the gas jet and electrostatic actuator was within 8% in the range from 35 to 1000 nN.

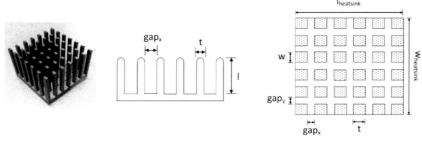

FIGURE 8.26

A fin-type electrostatic force calibrator [29].

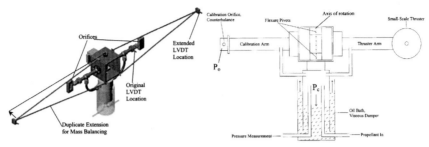

FIGURE 8.27

A gas-jet calibrator [30].

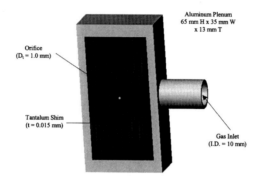

FIGURE 8.28

Orifice of gas jet calibrator [30].

8.1.6.2 Impulsive thrust

Reference thrust impulse is provided with various methods: impact hammer and electromagnetic and electrostatic actuators. A pendulum-type hammer is an option, and the magnitudes are adjustable with the initial height and hammer mass [25,32]. The simple structure and principle enable easy design, but some attention is necessary. The hammer may be bounced back by impengement so that collisions are induced many times. In such a case, attaching some soft materials to the impact point is effective to prevent the repetitive impact.

A velocimeter can determined the impact produced by the hammers, as shown in Fig. 8.29. Haag used a velocimeter, which is a sort of the photo-interrupter, moving and stationary rulings interrupt the incandescent-lamp light, and change in intensity is measured with a photo-interrupter so that velocity is evaluated using intensity histories [25]. High-speed cameras can also be used instead of the photo-interrupter. However, this may be an elaborating task because videos contain much unnecessary information.

The piezoelectric load cell can trace the time history of the reference impact because of higher stiffness than strain-gauge types owing to the higher elastic modulus of piezoelectric materials so that 10 kHz force change can be evaluated.

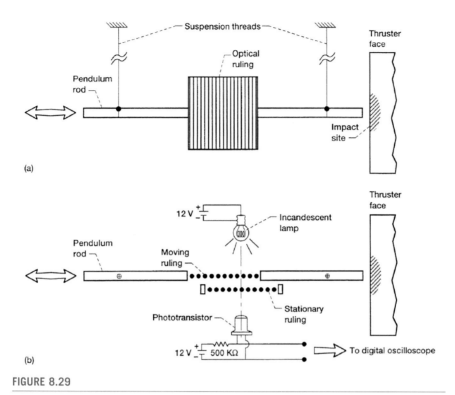

FIGURE 8.29

An impact hammer type calibrator with velocity meter [25].

Integration of instant force determines the reference impulse. The load cell can be used as a hammer, as shown in Figs. 8.30 and 8.31. The impulse bit (impulsive thrust) of the magnetoplasma dynamic thruster and pulsed plasma thruster was measured with the thrust stand, which was calibrated using an impact hammer with a load cell [33,34].

For simplicity, the pre-impact velocity is calculated using mechanical energy conservation. When the weight for impact is so small that the drag force due to motion in the air is negligible, the velocity is readily calculated. In the method, the adhesion is placed on the thrust stand to capture the weight after the impact, and this eliminates the necessity of determining post-impact velocity.

Electromagnetic actuators depicted in Figs. 8.24 and 8.25 can also provide impulsive force. The magnitude of the reference is adjustable with driving current and pulse width. The switching is readily implemented using a typical switching circuit using a field-effect transistor. Fig. 8.32 shows a time history of reference impulse at a turn-on period of 1 ms. The squared voltage pulse was applied at the time origin, and the reference thrust was varied in a triangle-like way owing to the inductance and resistor of the solenoid. The calibrator provided $5-100\ \mu$Ns thrust impulse.

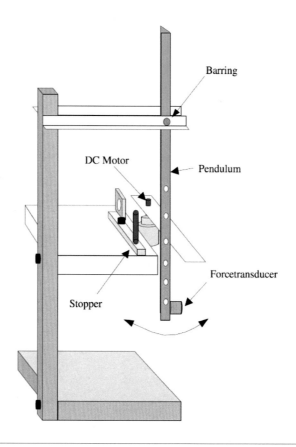

FIGURE 8.30

An impact hammer calibrator with load cell [34].

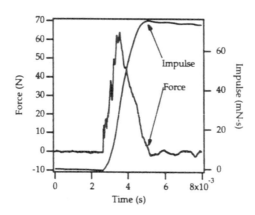

FIGURE 8.31

Time history of reference thrust for pendulum hammer calibrator with load cell [33].

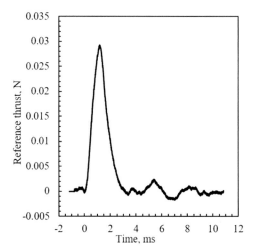

FIGURE 8.32

An example of reference thrust history produced by an electromagnetic actuator.

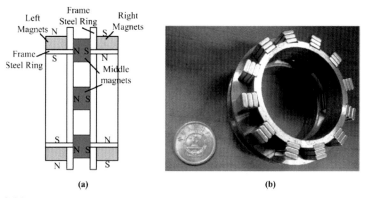

FIGURE 8.33

An electromagnetic actuator with a modified magnet layout [35].

Lam et al. applied the VCM calibrator presented in Fig. 8.25 to the calibration of impulsive thrust [27]. The calibrator provided impulsive reference thrust in the range from 12 to 668 μNs to show that the prototype thrust stand had an uncertainty of 11.38%. Tang et al. designed a new calibrator with a modified magnet layout to uniform the magnetic field [35]. Fig. 8.33 depicts the designed calibrator. The calibrator was calibrated using a mass balance and yielded an uncertainty of 10 μNs for 95% credibility.

Wong et al. applied the electromagnetic calibrator to the repetitively-firing pulsed thruster to evaluate the average thrust [36]. A train of pulsed firing produces quasi-steady displacement with periodic variation though a single impulsive thrust usually induces the oscillation expressed as Eq. 8.10. In tests, an electromagnetic actuator repeatedly provided mNs-class reference impulsive thrust 50 ms in width. Average deflection was proportional to both repetitive frequency and reference impulsive thrust. Hence, the test showed that the averaged thrust yielded by repetitively-firing pulsed thruster was evaluable using displacement.

Electrostatic actuators can also be applied to impulsive-thrust calibration. The calibrator illustrated in Fig. 8.26 produced pulsed reference thrust, either. Cheah et al. applied the fin-type calibrator (Fig. 8.26) to impulsive thrust calibration [29]. The calibrator produced impulsive thrusts of 7–340 μNs by the application of precisely controlled high-voltage pulses. The error is less than 5% above 25 μNs impulsive thrust. Gamero-Castaño prototyped μN thrust stand, calibrated using parallel plate electrodes, and showed that the thrust stand could evaluate thrust ranging from 3.45 to 39.3 μNs with excitation intervals between 20 and 80 ms [12]. Pancotti et al. examined the accuracy of the electrostatic calibrator in impulsive-thrust measurement using the impact hammer [37].

Note that attaching something such as coils or magnets to the load cell deteriorates the natural frequency. In such a case, the actual natural frequency should be calculated using the masses of attached objects and the spring constant of the load cell, data of which some manufacturers release. Moreover, the piezoelectric load cell is not useable to the steady-thrust calibration because of the leak current; impulse width is limited to approx. 1 s or less since the typical time constant of the electric discharge is less than 10 s. In contrast, strain gauge load cell is applicable to steady thrust calibration but is not appropriate for the impulse thrust calibration owing to lower stiffness than the piezoelectric type.

8.1.7 State-of-the-art

This chapter has dealt with the theory and practice of cutting-edge activities. Many works have been dedicated to increasing accuracy and extending thrust measurable range to enable reliable and convenient devices. Despite the great success in enhancing performance, various endeavors are still necessary to figure out the thruster characteristics. The section will introduce state-of-the-art thrust measurement: sub-μN thrust, specific impulse, thrust variation, and thrust vector.

8.1.7.1 Sub μN and μNs measurement

Especially for electric propulsion, which generally produced thrust on the orders of mN or sub-mN, sincere efforts have been dedicated to thrust measurement. Attraction to microthrusters has been revived owing to the emergence of nanosatellite-based missions, and the trend accelerated the demand for μN thrust evaluation. Such a level of thrust was determined using the displacement [26,38–40] and null-balance methods [16]. Some literature discussed nN-class resolution for μN thrust measurement. Today, some thruster produces thrust as low as sub-μN(μNs) level, and some research groups designed and examined the thrust stands that enable μN-class performance evaluation. Then, the section introduces some researches for sub-μN(μNs) thrust measurement.

A thrust target was applied to sub μN (μNs) thrust evaluation [41]. As shown in Fig. 8.34, a sensing plate (a thin aluminum plate 130 μm thick) was connected to MEMS-based capacitive force sensors (100-μN rated force, 5-nN resolution, and 50-N/m spring constant). The thruster exhaust jets impinged with the sensing plate so that the force was transferred to the force sensor. The output was monitored with a lock-in amplifier whose operational frequency lied between the facility-vibration frequency and the sensing-plate natural frequency to fine the resolution. The thrust target evaluated a μN-class thrust by a cold gas jet. The standard deviations of the thrust are 10 and 20 nN for reference thrusts below and above 30 μN, respectively. An electrospray device below 3 μN in thrust was also tested using the thrust target, and the tests showed that the thrust target was detectable thrusts below 50 nN.

Jamison et al. designed and tested a horizontal pendulum type thrust stand with an oil damper [30]. Fig. 8.27 depicts the schematic with the gas-jet calibrator, which was addressed in Section 8.1.6.2. Below 1 μN, an extension was added to the thrust stand arm 25 cm long from the center of rotation to enlarge the displacement. The displacement was evaluated with an LVDT 260 mV/mm in sensitivity. Gas-jet calibration showed that the stand measured thrust ranging from 86.2 to 712 nN, presenting errors of 10.7% and 2% at reference thrusts of 86.2 and 712 nN, respectively.

D'Souza et al. designed a horizontal-pendulum thrust stand with an oil damper and LVDT displacement sensor 1453 V/m in sensitivity [42]. The pendulum arm oscillation period T is 2.49 s. A pulse width of the reference impulsive thrust τ was varied such that $\tau/T = 10^0 \sim 10^{-5}$. The electrostatic calibrator was driven with time-varying reference (sudden increase with long-tail decay) and square pulse, as explained in Section 8.1.6.2. When the electrostatic calibrator provided reference impulsive thrust ranging from 7.25 nNs to 24.2 μNs, the thrust stand showed accurate output for $\tau/T \leq 0.1$, whereas presenting unacceptable errors for $\tau/T > 0.1$. Hence, the pulse width of the thruster firing should be shorter than 1/10 of the pendulum period. The stand evaluated 7 nNs with the worst error of 4.5% and showed linearity up to hundreds of μNs.

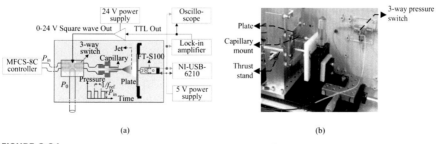

(a)   (b)

FIGURE 8.34

A thrust target using plate deflection [41].

8.1.7.2 Simultaneous evaluation of thrust and propellant consumption for solid propellant thruster

Some groups proposed the thrust stands that simultaneously evaluate the thrust and propellant consumption for solid propellant thrusters. In solid propellant thrusters, which retain the propellant inside, instantaneous propellant consumption has difficulty in measurement because requiring a sort of two-dimensional force vector measurement. Hence, averaged propellant rates, which are determined using mass difference before and after firing, have been conventionally used for performance evaluation. However, real-time or near real-time consumption rates are necessary to gain further insights into thrusters such as instantaneous specific impulses.

Hence, a vertical pendulum that can measure both impulse and thruster weight was developed [43]. The thrust stand rotates vertically so that the pendulum receives thruster weight and impulsive thrust. The propellant consumption can be measured using the offset of the equilibrium point of the balance, and the impulsive thrust can be evaluated using the amplitude of displacement because the thrust stand arm moves as Eq. (8.10) even when the equilibrium point is drifted owing to the variation in stationary force, i.e., the thruster weight. Tests using a laser-ablating thruster showed that the prototype evaluated the specific impulse on the order of 200 s with corresponding propellant consumption per shot of approx. 15 μg and impulsive thrust of approx. 30 μNs.

Yoshikawa et al. developed a calibration method for the simultaneous evaluation of propellant consumption, impulsive thrust, and average thrust for repetitive firing of pulsed plasma thrusters (PPTs) [44]. For impulsive-thrust calibration, the known-mass weights were simultaneously projected and ejected to simulate the PPT impulse, as shown in Fig. 8.35. Moreover, as presented in Fig. 8.36, a servomotor rotated the disk and released a known mass from the thrust stand arm to evaluate propellant consumption, which was determined using the arm-displacement drift from the equilibrium. The step-like thrust was simulated by dropping known mass to the arm to evaluate average thrust during repetitive pulsed firing. A thrust stand was calibrated in the impulsive thrust and accumulated mass loss ranging from 0 to 800 μNs and from 100 to 450 mg, respectively. Thrust measurement of a PPT showed that 1000-shot-accumulated mass loss (propellant consumption) ranged from 50 to 80 mg with errors of 1%−17%, and impulsive thrust was lowered from approximately 550 to 300 μNs through 10000 shots.

8.1.7.3 High-frequency thrust variation

As mentioned in Sections 8.1.2 and 8.1.3, the conventional thrust stand has a limitation in measurable frequency due to the resonance. However, thrust sometimes has variation during stable firing. In the start and interruption of firing, the thrust rapidly changes and contains high-frequency components, as the Fourier transfer of the single square pulse shows, so that the thrust stand is unable to track the rise and fall of thrust production and resultantly evaluate delay. The thrust variation, including ignition delay, needs to be measured from the viewpoint of stability because, in general, the variation and delay of actuators impede control and sometimes destabilize the system.

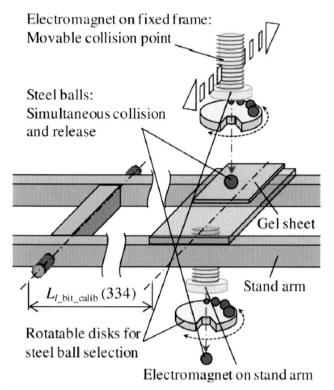

Electromagnet on fixed frame:
Movable collision point

Steel balls:
Simultaneous collision
and release

Gel sheet

$L_{l\_bit\_calib}$ (334)

Stand arm

Rotatable disks for
steel ball selection

Electromagnet on stand arm

FIGURE 8.35

A thrust stand and micromass balance for specific impulse measurement (Impulsive thrust calibration) [44].

Ahlfeld applied an augmented state-space Kalman filter to deconvolute thrust from a thrust stand arm response containing electric noises to evaluate 10-N class liquid propellant thruster [45]. Numerical simulation was conducted using a thrust stand model 150 Hz in natural frequency, which yields underdamped oscillation. The augmented deconvolute Kalman filter successfully deconvoluted the thrust variation 2 ms in rise/fall time and 30 ms in on/off time.

For lower thrust, acceleration measurement is applied to the null-balance thrust stand to compensate for the error due to the resonance, as shown in Fig. 8.37 [23,46]. In the null-balance method, the appropriate setting of the controller allows $c\dot{\theta} + k\theta \approx 0$ in Eq. (8.11), and hence, the thrust is evaluated using $\ddot{\theta}$ and actuator-driving current i. The application of an accelerometer extended the maximum measurable frequency from 3 Hz to 80 Hz if a 10% error is accepted. The frequency limit was attributable to the transnational flexibility of the flexural pivot. As previously mentioned in Section 8.1.5.2, the flexure hinges had transnational flexibility (Fig. 8.17). The pendulum (arm, thruster, and counterweight) mass and the flexibility form the mass-spring system, which has a 150 Hz class natural frequency.

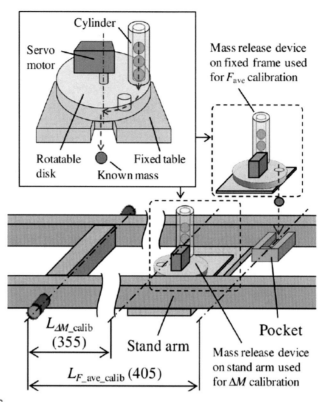

FIGURE 8.36

A thrust stand and micromass balance for specific impulse measurement (Propellant-consumption and average-thrust calibration) [44].

The translational resonance deteriorated the frequency response. Hence, the authors proposed to execute a two-point acceleration measurement to evaluate both rotational and translational accelerations under the condition where the C.G. is set on the hinge (rotational axis) [47]. Thrust variation up to 350 Hz except 20−180 Hz was successfully evaluated, though the thrust stand yielded 50% errors in the 20−180 Hz range.

8.1.7.4 Thrust vector

The thrust vector has been evaluated using some thrust stands. Determinating side thrusts and torques is an elaborating work because of their weakness compared with the main axis thrust. Hence, the plume was sometimes diagnosed to calculate the thrust vector [48]. Kami and Herron designed and tested a mercury-floated type to evaluate the thrust vector of ion thrusters [49]. Efforts on the thrust vector evaluation are maintained despite fewer researches than those on main-axis thrust evaluation.

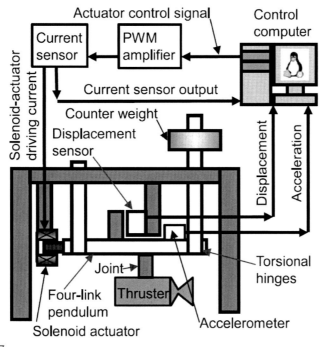

FIGURE 8.37

A null-balance thrust stand with one-point acceleration measurement for thrust variation measurement [46].

Three-component thrust vector measurement of milli-Newton class thrust was conducted by Hughes and Oldfield using the pendulum shown in Fig. 8.38 [50]. This is a triple pendulum, the axis of which is perpendicular to each other. One of the arms formed an inverted vertical pendulum to evaluate the main thrust. The displacements of the three arms were nulled by the control, and the thrust vector was evaluated using the voltage applied to the actuator. Tests showed that the thrust stand evaluated thrust ranging from 0 to 13 mN with a resolution below 10 μN and main and orthogonal accuracy of 1.2% and <1% of full scale.

Nagao et al. adapted the null-balance method to two-component thrust vector measurement for 30- mN class Hall thrusters [14] (Fig. 8.39). The thrust stand had two knife-edged hinges and formed a dual pendulum. The pendulum position was zeroed by the nulling-control using simple VCMs, and the two-component thrust vector was evaluated using VCM-driving currents. Calibration showed that the main and transverse components of the thrust vector were linear with VCM currents with errors of 0.25 mN (2.1%) and 0.09 mN (1.4%), respectively. A 0.5-kW class Hall thruster 20 mN in thrust was tested using the thrust stand, and the test showed that the thrust vector was steerable by distribution adjustment of propellant mass flux. The prototype thrust stand evaluated both 17-mN main

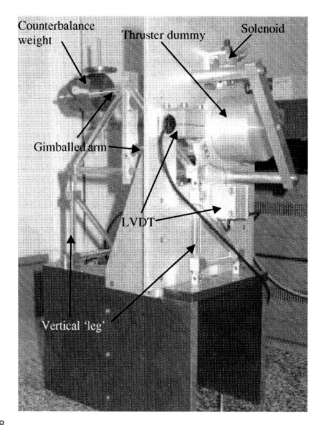

FIGURE 8.38

A three-component null-balance thrust stand [50].

and 1-mN transverse components of thrust and showed a thrust vector angle of 2.3°
with ±0.2° error.

A magnetically levitating thrust stand was developed to evaluate the six-
component thrust vector of a 0.5 N-class thruster by Kakami [17]. Multicomponent
measurement complicates the structure because it necessitates at least six pendulum
arms and hinges. Hence, magnetic levitation was introduced to eliminate the hinge
and pendulum arms. Fig. 8.40 depicts a schematic of the thrust stand. The triangle
levitator, on which a thruster is mounted, was magnetically supported with six
VCMs to levitate it at the target position. The test showed that the null-balance control levitated the levitator at the target. The residual displacements are on the order of
nm with variations up to 2 μm (rms). With regard to angle, the residual error is on the
order of 10^{-3} mrad and 10^{-2} mrad (rms), respectively. The thrust stand was calibrated using the calibrator presented in Fig. 8.25, and that the VCM driving currents
are proportional to thrust components. Then, the relationship between thrust and
VCM driving currents was evaluated in a matrix form with a coefficient of

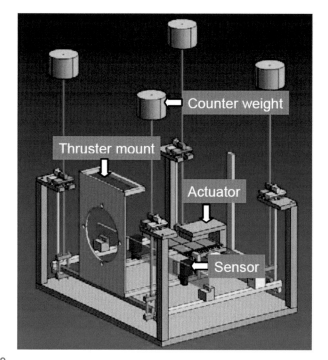

FIGURE 8.39

A two-component null-balance thrust stand [14].

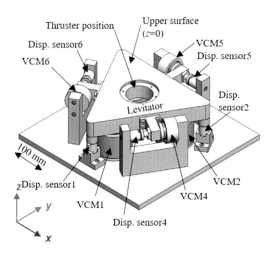

FIGURE 8.40

A six-component magnetic-levitation thrust stand [17] device development.

determination of 0.98. Dynamic response was determined by sinusoidally varying the calibrator driving current. The maximum measurable frequencies are 3 and 0.12 Hz for the main thrust (z-axis in Fig. 8.40) and x-axis torque.

8.1.8 Summary

This chapter addressed the thrust measurement and calibration for micropropulsion with the analysis and implementation. From the dawn of the space age, enormous endeavors have been devoted to augmenting accuracy and expanding measurable thrust range. Various calibration devices were also examined since calibration requires careful work; the human approaching the thrust stand induces mechanical noise. Sincere devotion has successfully enhanced accuracy and extended measurable range. Most readers would not doubt the proverb *Scientia est potentia*, and the thrust stands have provided indispensable *scientia* in propulsion.

However, most conventional thrust stands have given the time-averaged main-axis thrust, which is usually larger than μN order, whereas there are various hidden characteristics such as vector, time variation, and small thrust. Moreover, knowledge, experience, and elaborate works are still mandatory for thrust measurement, though concerning the role, the thrust stand is essentially the same as the household gravimetric gauges, which are manageable even for children to use. Hence, continuous effort is necessary to extend the measurable thrust level to nN, expand the frequency range in variable thrust to kHz, enable six-component thrust vector evaluation, and enhance usability. The author believes that further tenacious endeavors on thrust measurement would make a significant contribution to space development because unveiling all the hidden figures can lead to the success of novel and prominent propulsion systems.

References

[1] Froelich, R., Papapoff, H., 1959. Reaction wheel attitude control for space vehicles. IRE Trans. Automatic Control 4 (3), 139–149.
[2] Beachley, N.H., 1963. A Single-Axis Space Simulator for Testing the OGO Altitude Control System. Technical Report STL-2313-6007-KU-000, NASA-CR-77223. NASA.
[3] Smith, J.D., 1966. Technical Report NASA-CR-72021. NASA.
[4] Sidney, Z., Kemp, R.F., 1969. Colloid microthruster test stand. J. Spacecraft Rockets 6 (10), 1144–1147.
[5] Ferrara, J., Larocca, A., Malherbe, P., 1968. Microthrust and impulse measurement techniques for electrical thrusters. In: 4th Propulsion Joint Specialist Conference.
[6] Cutler, W., 1968. Development of a microthrust stand for direct thrust measurement. In: 4th Propulsion Joint Specialist Conference.
[7] Stark, K.W., 1971. Micro-pound Extended Range Thrust Stand. US3572104A, US Patent.

[8] Stark, K.W., Dennis, T., McHugh, D., Williams, T., 1971. Design and Development of a Micropound Extended Range Thrust Stand (MERTS). Technical Report NASA-TN-D-7029. NASA.

[9] Crossway, F.L., Kalb, H.T., 1968. Evaluation of a Static and Dynamic Rocket Thrust Measurement Technique. Technical Report AEDC-TR-68-117. Arnold Engineering Development Center, U.S. Air Force.

[10] Foote, J.F., 1969. Final report. Technical Report NASA-CR-101932. Apollo SM-LM RCS Engine Development Program Summary Report, vol. 4. NASA.

[11] Haag, T.W., 1995. PPT Thrust Stand. Technical Report NASA-TM-107066. NASA.

[12] Gamero-Castaño, M., 2003. A torsional balance for the characterization of micronewton thrusters. Rev. Sci. Instrum. 74 (10), 4509–4514.

[13] Orieux, S., Rossi, C., Est'eve, D., 2002. Thrust stand for ground tests of solid propellant microthrusters. Rev. Sci. Instrum. 73 (7), 2694–2698.

[14] Nagao, N., Yokota, S., Komurasaki, K., Arakawa, Y., 2007. Development of a two-dimensional dual pendulum thrust stand for hall thrusters. Rev. Sci. Instrum. 78 (11), 115108.

[15] Xu, K.G., Walker, M.L.R., 2009. High-power, null-type, inverted pendulum thrust stand. Rev. Sci. Instrum. 80 (5), 055103.

[16] Rocca, S., Menon, C., Nicolini, D., 2006. FEEP micro-thrust balance characterization and testing. Meas. Sci. Technol. 17 (4), 711–718.

[17] Kakami, A., Hanyu, K., Yano, Y., 2020. Magnetically-levitated thrust stand for evaluating 6-component thrust vector of 1-n class onboard propulsion devices. Aero. Sci. Technol. 104, 105896.

[18] Kakami, A., Muto, T., Yano, Y., Tachibana, T., 2015. A method for evaluating the thrust of a space propulsion device with wide range time variations using a disturbance observer. Rev. Sci. Instrum. 86 (11), 115114.

[19] Yanagi, R., Kimura, I., 1982. New type of target for the measurement of impulse bits of pulsed plasma thrusters. J. Spacecraft Rockets 19 (3), 246–249.

[20] Kuwahara, D., Koyama, Y., Otsuka, S., Ishii, T., Ishii, H., Fujitsuka, H., Waseda, S., Shinohara, S., 2014. Development of direct thrust measurement system for the completely electrodeless helicon plasma thruster. Plasma Fusion Res. 9, 3406025.

[21] Longmier, B.W., Reid, B.M., Gallimore, A.D., Chang-Diaz, F.R., Squire, J.P., Glover, T.W., Chavers, G., Bering, E.A., 2009. Validating a plasma momentum flux sensor to an inverted pendulum thrust stand. J. Propul. Power 25 (3), 746–752.

[22] Kokal, U., Celik, M., 2017. Development of a mili-Newton level thrust stand for thrust measurements of electric propulsion systems. In: 2017 8th International Conference on Recent Advances in Space Technologies (RAST), pp. 31–37.

[23] Kakami, A., Kashihara, K., Takeshida, S., Yano, Y., 2016. A new thrust measurement method for evaluating higher frequency variation by applying acceleration measurement to null-balance method. Trans. Jpn. Soc. Aeronaut. Space Sci. 14 (ists30), Pb 123–Pb 130.

[24] Cubbin, E.A., Ziemer, J.K., Choueiri, E.Y., Jahn, R.G., 1997. Pulsed thrust measurements using laser interferometry. Rev. Sci. Instrum. 68 (6), 2339–2346.

[25] Haag, T.W., 1955. PPT Thrust Stand. Technical Report NASA-TM-107066. Arnold Engineering Development Center, U.S. Air Force.

[26] Nakagawa, Y., Tomita, D., Koizumi, H., Komurasaki, K., 2018. Design and test of a 100 μn-class thrust stand for a miniature water ion thruster with cubesat. Trans. Jpn. Soc. Aeronaut. Space Sci. 16 (7), 673–678.

[27] Jiang, K.L., Koay, S.C., Lim, C.H., Cheah, K.H., 2019. A voice coil based electromagnetic system for calibration of a sub-micronewton torsional thrust stand. Measurement 131, 597−604.

[28] Lun, J., Law, C., 2014. Direct thrust measurement stand with improved operation and force calibration technique for performance testing of pulsed micro-thrusters. Meas. Sci. Technol. 25 (9), 095009.

[29] Cheah, K.H., Low, K., Tran, Q., Lau, Z., 2015. Development of an electrostatic calibration system for a torsional micronewton thrust stand. IEEE Trans. Instrum. Meas. 64 (12), 3467−3475.

[30] Jamison, A.J., Ketsdever, A.D., Muntz, E.P., 2002. Gas dynamic calibration of a nano-newton thrust stand. Rev. Sci. Instrum. 73 (10), 3629−3637.

[31] Selden, N.P., Ketsdever, A.D., 2003. Comparison of force balance calibration techniques for the nano-Newton range. Rev. Sci. Instrum. 74 (12), 5249−5254.

[32] Domonkos, M., Gallimore, A., Myers, R., Thompson, E., 1995. Preliminary pulsed MPD thruster performance. In: 31st Joint Propulsion Conference and Exhibit.

[33] Cubbin, E.A., Ziemer, J.K., Choueiri, E.Y., Jahn, R.G., 1995. Laser interferometry for pulsed plasma thruster performance measurement. In: 1995 International Electric Propulsion Conference, Number IEPC-95-195.

[34] Kakami, A., 2003. Water Propellant Pulsed Plasma Thruster. PhD thesis. University of Tokyo.

[35] Tang, H., Shi, C., Zhang, X., Zhang, Z., Cheng, J., 2011. Pulsed thrust measurements using electromagnetic calibration techniques. Rev. Sci. Instrum. 82 (3), 035118.

[36] Wong, A.R., Toftul, A., Polzin, K.A., Boise Pearson, J., 2012. Non-contact thrust stand calibration method for repetitively pulsed electric thrusters. Rev. Sci. Instrum. 83 (2), 025103.

[37] Pancotti, A.P., Gilpin, M., Hilario, M.S., 2012. Comparison of electrostatic fins with piezoelectric impact hammer techniques to extend impulse calibration range of a torsional thrust stand. Rev. Sci. Instrum. 83 (3), 035109.

[38] Ziemer, J.K., 2001. Performance measurements using a sub-micronewton resolution thrust stand. In: 27th International Electric Propulsion Conference. number Paper IEPC-01-238, Pasadena, CA.

[39] Soni, J., Roy, S., 2013. Design and characterization of a nano-Newton resolution thrust stand. Rev. Sci. Instrum. 84 (9), 095103.

[40] Hathaway, G., 2015. Sub-micro-Newton resolution thrust balance. Rev. Sci. Instrum. 86 (10), 105116.

[41] Chakraborty, S., Courtney, D.G., Shea, H., 2015. A 10 nn resolution thrust-stand for micro-propulsion devices. Rev. Sci. Instrum. 86 (11), 115109.

[42] D'Souza, B.C., Ketsdever, A.D., 2005. Investigation of time-dependent forces on a nanonewton-second impulse balance. Rev. Sci. Instrum. 76 (1), 015105.

[43] Ketsdever, A.D., D'Souza, B.C., Lee, R.H., 2008. Thrust stand micromass balance for the direct measurement of specific impulse. J. Propul. Power 24 (6), 1376−1381.

[44] Yoshikawa, T., Tsukizaki, R., Kuninaka, H., 2018. Calibration methods for the simultaneous measurement of the impulse, mass loss, and average thrust of a pulsed plasma thruster. Rev. Sci. Instrum. 89 (9), 095103.

[45] Ahlfeld, R.B.H., 2014. Deconvolution Filters for Dynamic Rocket Thrust Measurements. Master's thesis. TU Delft.

[46] Kakami, A., Tachibana, T., 2013. Thrust evaluation in wide frequency range using active control and disturbance observer. J. Propul. Power 29 (6), 1274−1281.

[47] Yamauchi, Y., Yano, Y., Kakami, A., 2019. A horizontal-pendulum type thrust stand for evaluating higher frequency variation by applying acceleration measurement to null-balance method. Trans. Jpn. Soc. Aeronaut. Space Sci. 17 (2), 175−180.

[48] Pollard, J., Welle, R., 1995. Thrust vector measurements with the t5 ion engine. In: 31st Joint Propulsion Conference and Exhibit.

[49] Kami, S., Herron, B., 1970. An ion engine thrust vector measuring device. In: 8th Electric Propulsion Conference, Number AIAA Paper 70-1104.

[50] Hughes, B., Oldfield, S., 2004. Traceable calibration of the 3 axis thrust vector in the mn range. In: 4th International Spacecraft Propulsion Conference. ESA Special Publication, p. 128.1.

Nanoenergetic for micropropulsion

Ruiqi Shen, PhD [1,2], Yinghua Ye, PhD [1,2], Luigi T. DeLuca, PhD [1,3], Chengbo Ru, PhD [1,2], Xiaoyong Wang, MS [1,2], Zhang He, MS [1,2]

[1]*Institute of Space Propulsion, School of Chemical Engineering, Nanjing University of Science and Technology, Nanjing, Jiangsu, China;* [2]*The Key Laboratory of Nano-micro Energetic Devices of Ministry of Industry and Information Technology, Nanjing University of Science and Technology, Nanjing, Jiangsu, China;* [3]*Space Propulsion Laboratory, Department of Aerospace Science and Technology, Politecnico di Milano, Milano, Italy*

9.1 Introduction

The easy maintenance and the simple structure of solid propellant chemical thruster are more advantageous than other traditional propulsion techniques, such as electrical propulsions, cold gas propulsions, and mono-/bichemical propellant propulsions. The controllable and stable combustion of solid propellants for micropropulsion is an important requisite for microsatellite applications because it is required that microthrusters can turn on and turn off, if possible, can control thrust, and some MEMS thrusters need that the solid propellant burns below $300\ \mu m$ diameter of grain. Unfortunately, we know that the solid propellant grains used in microthruster, such as ammonium perchlorate hydroxyl-terminated polybutadiene (AP)-HTPB propellant, lead styphnate propellant, and nanothermite propellant, are not only difficult to stop burning once they are ignited, but also burn unsteadily, even remain unburned in small scale. The control of solid propellants combustion, stable burning of propellant, and switching on/off of thruster are major challenges for the microscale solid propellant thruster.

Some innovative technologies were studied and proposed to solve these problems, primarily based on physical augmentation of combustion and nanotechnology in Fig. 9.1. These developing technologies are laser augmented chemical propulsion [1,2], microwave augmented propulsion [3,4], and electrochemical propulsion [5,6], as well as microsolid propulsion using nanothermite propellant [7] and nanostructured energetic propellant [8].

9.2 Combustion equations of nanoenergetic propellant

Conventional solid propellants, such as double-base propellant, composite modified double-base propellant, and ammonium perchlorate (AP)-HTPB composite propellant,

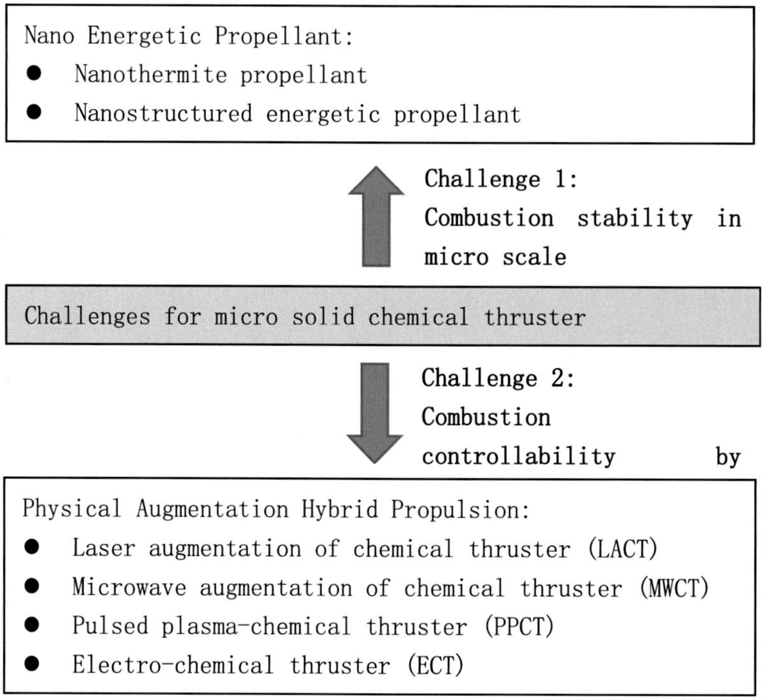

Nano Energetic Propellant:
- Nanothermite propellant
- Nanostructured energetic propellant

Challenge 1:
Combustion stability in micro scale

Challenges for micro solid chemical thruster

Challenge 2:
Combustion controllability   by

Physical Augmentation Hybrid Propulsion:
- Laser augmentation of chemical thruster (LACT)
- Microwave augmentation of chemical thruster (MWCT)
- Pulsed plasma-chemical thruster (PPCT)
- Electro-chemical thruster (ECT)

FIGURE 9.1

Solid propellant and its hybrid propulsion for challenges.

are widely used in solid rocket motor (SRM), but they are unsuited for microthruster applications because their critical diameter (self-extinction diameter for combustion at ambient pressure and temperature) is too large and burns unsteadily under low pressure and in small thruster chambers. In the early development, AP-HTPB propellants [9] and glycidyl azide polymer (GAP) propellants [10,11] were used in microthruster, but they were replaced by lead styphenate (LTNR) propellant [12] and nickel hydrazine nitrate (NHN) propellant [13] that have high burning rate, small critical diameter, and high thermal sensitivity. Because LTNR and NHN are highly sensitive energetic compounds and primer explosives used in pyrotechnic devices, such as igniter and detonator, they pose safety risk concerns to micropropulsion applications. Alternatively, solid propellants of low sensitivity, high burning rate, and small critical diameter are required for microthrusters.

The dependence of burning rate on the diameter of propellant is deduced from solving the heat balance equations to identify the design requirements of nanoenergetic propellant. A simplified heat balance equation consists of the heat of condensed phase from the initial temperature (T_0) to the burning surface temperature (T_s), the heat loss from propellant grain surface (q_{loss}), the combustion heat of propellant grain (q_{chem}) in condensed phase, and the irradiant heat (q_{rad}) is shown in and Eq. (9.1) and schematically presented in Fig. 9.2.

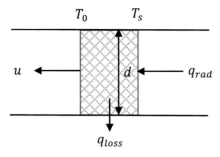

FIGURE 9.2

Combustion wave of micropropellant grain.

$$\rho Cu(T_s - T_0) + 4k\frac{1}{d}(T_s - T_0) = q_{chem} + q_{rad} \tag{9.1}$$

$$u = \frac{q_{chem}}{\rho C(T_s - T_0)} + \frac{q_{rad}}{\rho C(T_s - T_0)} - \frac{4k}{\rho C}\frac{1}{d} \tag{9.2}$$

$$u = \alpha + \beta q_{rad} - \gamma\frac{1}{d} \tag{9.3}$$

where ρ and C are the density and the heat capacity of propellant, k is the heat loss coefficient between the grain peripherical surface and the attached outside boundary, and d is the grain diameter.

If the burning rate of propellant is zero $(u = 0)$, the critical diameter of propellant is shown in Eq. (9.4), The critical diameter depends on combustion heat and combustion temperature. If gaseous phase reactions are neglected, the combustion temperature is the temperature at the burning surface, it can even be the final decomposition temperature of the propellant.

$$q_{cr} = \frac{4k(T_s - T_0)}{q_{chem} + q_{rad}} \tag{9.4}$$

The burning rate is lower in small grain diameter than in larger diameter, and when the grain diameter approaches the critical diameter, the grain does not burn normally. High burning rate (or low burning temperature/decomposition temperature) and high combustion heat benefit the burning process on a small scale, as well as external radiation heat, such as that from a laser beam or microwave field.

The aforementioned discussion in previous works suggests that the selecting principles of solid propellant for micropropulsion applications are high burning rate and high reaction heat, in which primer explosive propellants and nanothermite propellants are the best candidates.

9.3 Interior ballistic equations of microthruster

The propulsion performance of thermites can be calculated using thermochemistry and rocket equilibrium equations. The internal ballistic equations of a solid chemical microthruster based on the principle of SRM are given as follows:

$$\text{Thrust: } F = \dot{m}v_e + A_e(P_e - P_a) \tag{9.5}$$

$$\text{Impulse : } I = \int_0^{t_b} F dt \tag{9.6}$$

$$\text{Specific Impulse : } I_{sp} = I/(m \cdot g_0), \tag{9.7}$$

$$I_{sp} = \left\{ 2\frac{RT_c}{M}\frac{k}{k-1}\left[1 - (P_e - P_c)^{\frac{k-1}{k}}\right] \right\}^{\frac{1}{2}}, \text{(chem. frozen flow)} \tag{9.8}$$

$$\text{Characteristic velocity: } C^* = P_c A_t/\dot{m}, \ C^* = \sqrt{nRT_c}/\sqrt{k\left(\frac{2}{k+1}\right)^{(k+1)/(k-1)}} \tag{9.9}$$

where $\dot{m}$ is the mass rate from the nozzle, v_e, A_e, and P_e are the gas flow velocity, the cross-sectional area of nozzle and the pressure at the nozzle exit, P_a is the pressure of the environment, P_c and T_c are the pressure and the temperature in the combustion chamber, m is the mass of propellant, M is the relative molecular mass of gaseous products, A_t is the nozzle throat area, k is the adiabatic exponent of gas products $(k = c_p/c_v)$, and R is the gas constant.

The theoretical specific impulse depends on the combustion heat of propellant, and the relationship is shown in Eq. (9.10)

$$I_{sp} = \sqrt{2\eta H}, \ \eta = 1 - (1/\varepsilon)^{(k-1)/k} \tag{9.10}$$

where H is the combustion heat, η is the combustion efficiency of propellant, and ε is the pressure expansion ratio $(\varepsilon = P_c/P_e)$.

The theoretical specific impulse relates to the average molecular mass of gas products and the combustion heat or the combustion temperature,

$$I_{sp} \sim \sqrt{T_{max}/M} \tag{9.11}$$

where T_{max} is the maximum temperature of combustion products.

The specific impulse relates to the environmental pressure, which is higher in vacuum than at sea level. The vacuum specific impulse is shown in Eq. (9.12)

$$I_{spv} = I_{sp} + P_e A_e/\dot{m} \tag{9.12}$$

where I_{spv} is the vacuum-specific impulse $(P_a = 0)$.

9.4 Microthrust balance

Because very small amount (in 10−100 mg) of nanoenergetic propellant is loaded into a microthruster, it is difficult to test its burning rate using techniques similar to those implemented for conventional solid propellants. However, microthrust balances are useful to test the combustion performances of nanoenergetic propellants. Three kinds of thrust balances, vertical thrust balance [14,15], horizontal thrust balance [16], and nulled thrust balance [17], are often used to measure the thrust and impulse of thruster, in which the vertical thrust balance based on the pendulum principle is easily affected by gravity, but the horizontal thrust balance and the nulled thrust balance can separate the thrust from the weight of thruster, making the two balances particularly suitable for measuring small thrust generated by heavy thrusters. The vertical thrust balance and the horizontal thrust balance are used to test thrust from μN to mN in MIIT Key Laboratory of nanomicroenergetic devices at Nanjing University of Science and Technology.

9.4.1 Vertical thrust balance

The VTB is a simple device (shown in Fig. 9.3) that can measure thrust from μN to mN to evaluate the propulsion abilities of microthruster using the pendulum principle. The displacement of the thruster holder at the bottom of the balance arm is measured by a Hall magnetic sensor, which detects the magnetic change from the magnetic field of a magnetic element attached to the thruster holder. The burning time of propellants is detected by a photodetector. Evaluating the performance of nanoenergetic propellant, a simple thruster is designed and fabricated, in which the thruster is only a bore on a substrate of silicon, polymer, or glass and the propellant is loaded into the bore by a vacuum drawing device. The nanoenergetic propellant in the simple thruster is ignited by a free-running Nd:YAG laser beam output from a coupled optic fiber (Figs. 9.3 and 9.4).

The change of displacement along the arc-shaped cruise from the balance position of the thruster to a new position at the time,

$$L=\frac{1}{2}at^2=\frac{Ft^2s_1^2}{2J} \tag{9.13}$$

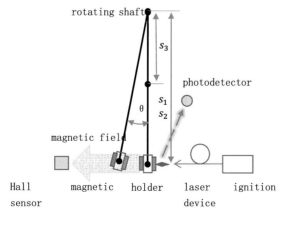

FIGURE 9.3

Vertical thrust balance (VTB) measurement principle of thrust and burning time from microthruster.

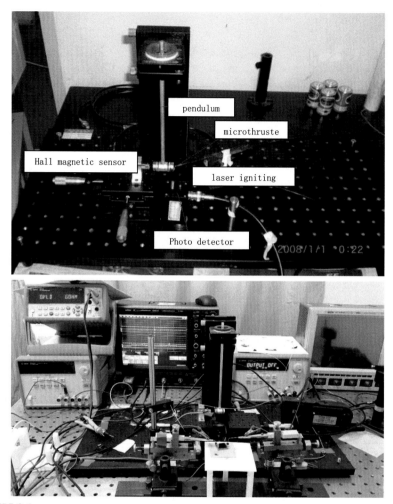

FIGURE 9.4

Vertical thrust balance (VTB) measurement set of thrust and burning rate from microthruster.

and

$$a = a_0 \cdot s_1$$

$$a_0 = \frac{M}{J} = \frac{F s_1}{J}$$

where a, a_0, and t are the linear acceleration, the angular acceleration, and the displacement time of thruster, F is the thrust, s_1 is the distance from the center of mass of the thruster holder to the rotating shaft, M is the torque of balance, and J is the rotational inertia.

The rotational inertia consists of the rotational inertia of beam arm, thruster holder, and thruster,

$$J = m_1 s_1^2 + m_2 s_2^2 + m_3 s_3^2 \qquad (9.14)$$

where s_1, s_2, and s_3 are the distances from the center of mass of the thruster holder, thruster, and balance arm, m_1, m_2, and m_3 are the masses of the thruster holder, thruster, and balance arm.

The rotational angle of balance (θ) is shown in Eq. (9.15).

$$\theta = arctg \frac{dx/s_1}{\sqrt{1 - \left(\dfrac{dx}{ds_1}\right)^2}} \qquad (9.15)$$

where dx is the horizontal displacement distance from the balance position of the thruster to position at time, and θ is the rotational angle of balance.

The displacement of the thruster is shown in Eq. (9.16)

$$L = s_2 \theta \qquad (9.16)$$

where s_2 is the distance from the center of mass of the thruster to the rotating shaft, and θ is the rotational angle of balance.

The thrust can be calculated by the conservation equation of work and potential energy in Eq. (9.17)

$$F \cdot L = mg \cdot dh \qquad (9.17)$$

and

$$F \cdot L = (m_1 s_1 + m_2 s_2 + m_3 s_3) \cdot g \cdot (1 - \cos(\theta))$$

where dh is the vertical distance of mass center from the initial position to "t" position.

The thrust can be calculated by Eq. (9.18),

$$F = \frac{\sqrt{2(m_1 s_1 + m_2 s_2 + m_3 s_3) \cdot g \cdot (1 - \cos(\theta)) \cdot J}}{t s_1} \qquad (9.18)$$

9.4.2 Horizontal thrust balance

The horizontal thrust balance (HTB) is better than the vertical thrust balance because it can separate thrust from weight, and it is only affected weakly by gravity and thus able to measure small thrust from heavy thruster accurately. The top view of HTB is shown in Fig. 9.5. A counterweight balances the center mass of HTB (including thruster) to a flexible shaft position. The displacement (or calculated rotation angle) of the balance arm pushed by the thrust of the tested sample is measured by a laser displacement sensor. The deflection of the flexible shaft and the damping coefficient of the electromagnetic damper decide the measuring parameters of accuracy, range, and frequency response. The actual setup is shown in Fig. 9.6.

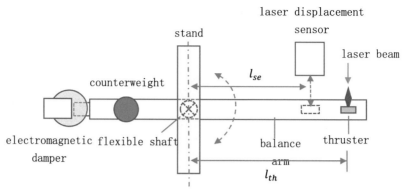

FIGURE 9.5

Top view of horizontal thrust balance.

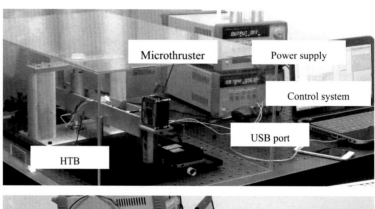

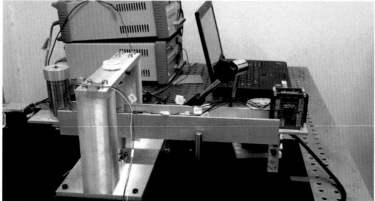

FIGURE 9.6

Experimental setup of horizontal thrust balance.

The dynamic model of the torsional pendulum is a second-order system of the mass-spring-damp model.

$$J\frac{d^2\theta}{dt^2} + c\frac{d\theta}{dt} + k\theta = l_{th}F(t) \tag{9.19}$$

where θ is the rotation angle of the torsional pendulum, J is the moment of inertia to the flexible shaft, c is the damping coefficient of electromagnetic damper, k is the torsional stiffness coefficient of the flexible shaft, l_{th} is the distance from the axis of the flexible shaft to the force application point of thrust.

The rotation angle of the torsional pendulum can be converted into the displacement of the balance arm (measured by a laser displacement sensor).

$$tg(\theta) = \frac{x}{l_{se}}, \quad \left(\theta = \frac{x}{l_{se}}, \ \ if \ \theta \ll 1\right) \tag{9.20}$$

where x is the displacement measured by the sensor, l_{se} is the distance from the axis of the flexible shaft to the cross point of the vertical line of the sensor and balance arm.

The dynamic equation of displacement is shown in Eq. (9.21),

$$\frac{d^2x}{dt^2} + 2\mu\frac{dx}{dt} + \omega_0^2 x = \frac{l_{se}l_{th}}{J}F(t) \tag{9.21}$$

where ω_0 is the natural-vibration frequency (Hz) ($\omega_0 = \sqrt{k/J}$), and μ is the viscosity coefficient ($\mu = c/2J$).

The solution of Eq. (9.21) at $x(0) = 0$ and $\dot{x}(0) = 0$,

$$x(t) = Aexp(-\mu t)\sin(\widehat{\omega}t) \tag{9.22}$$

$$A = \frac{l_{se}l_{th}}{J\widehat{\omega}}I, \quad \widehat{\omega} = \sqrt{\omega_0^2 - \mu^2} \tag{9.23}$$

where A is the maximum disposition amplitude measured by displacement sensor, and I is the impulse of the tested thruster.

If the damping of HTB is neglected ($\mu = 0$),

$$x(t) = A_0\sin(\omega_0 t), \quad A_0 = \frac{l_{se}l_{th}}{J\omega_0}I \tag{9.24}$$

The impulse of tested thruster by undamped HTB is

$$I = KA_0, \quad K = \frac{J\omega_0}{l_{se}l_{th}} \tag{9.25}$$

If $\mu \neq 0$, the impulse of the tested thruster is shown in Eq. (9.26)

$$I = \frac{J\omega_0\sqrt{1-\varepsilon^2}}{l_{se}l_{th}}A, \quad or \quad I = K\sqrt{1-\varepsilon^2}A \tag{9.26}$$

where ε is the damping ratio of HTB, $\varepsilon = \mu/\omega_0$.

Eqs. (9.25) and (9.26) indicate that there was a relationship between the impulse of the thruster and the amplitude of displacement. After calibrating K value or ω_0 and J, the impulse can be measured by measuring the amplitude of displacement (A or A_0).

ω_0 and J can be calibrated by direct method and indirect method in the experiment, in which a given thrust or impulse is used in direct calibration, while the parallel axis theorem is used in the indirect calibration. Providing a high accuracy thrust or impulse source is very difficult, so the parallel axis theorem is a suitable method.

In the parallel axis theorem, if a body whose mass is m rotates around the central axis of mass, its moment of inertia is J_0, and when the body shifts to another position from the central axis, the moment of inertia of the shifted rotation system will be given in Eq. (9.27).

$$J = J_0 + mx^2 \qquad (9.27)$$

where x is the shift distance of the body from the central axis of mass to a new position, m is the mass of the body, J_0 is the moment of inertia rotating around the central axis of mass.

The parallel axis theorem methods using a pair of rotational symmetry weights are applied to calibrate the moment of inertia of HTB, as seen in Fig. 9.7.

Two identical weights are used to calibrate the moment of inertia, and the moment of inertia of one weight is as follows:

$$J_w = \frac{1}{8}mD^2 \qquad (9.28)$$

where D is the diameter of the cylinder, m is the mass of calibrating cylinder.

When a pair of identical weights are placed at the axisymmetric position from the axis, the sum of moment of inertia of balance and a pair of weights is given by

$$J = J_0 + 2\left(J_w + md^2\right) \qquad (9.29)$$

where J_0 is the moment of inertia of HTB, and d is the distance from the center mass of calibrating weight to the axis.

As the period of rotation of HTB is

$$T = \frac{2\pi}{\omega} = 2\pi\sqrt{J/k} \qquad (9.30)$$

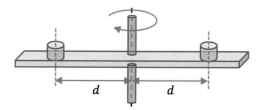

FIGURE 9.7

Calibration of the moment of inertia of HTB by a pair of weights.

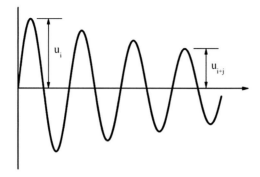

FIGURE 9.8

Oscillating wave of disposition in the process of calibration.

and

$$\frac{T_0^2}{T^2} = \frac{J_0}{J} = \frac{J_0}{J_0 + 2\left(J_w + md^2\right)}$$

according to Eq. (9.30), the moment of inertia of HTB is given by

$$J_0 = \frac{2T_0^2}{T^2 - T_0^2}\left(J_w + md^2\right) \tag{9.31}$$

Changing the position of calibrating weight, a series of oscillating waves of disposition (Fig. 9.8) can be recorded to calculate the damping ratio of HTB (ε), the natural-vibration frequency (ω_0), and the moment of inertia (J_0). The impulse of the thruster can be calculated by Eq. (9.26) using the calibrated parameters.

The natural-vibration frequency (ω_0) can be calculated by an experimental average period of displacement signal in ahead of 20 cycles,

$$\omega_0 = \frac{2\pi}{T} \tag{9.32}$$

The damping ratio of HTB (ε) also can be calculated by the ith and $i + j$th peak amplitudes.

$$\varepsilon = \frac{1}{2\pi j}\ln\left(\frac{u_i}{u_{i+j}}\right) \tag{9.33}$$

where u_i and u_{i+j} are the ith and $i + j$th peak amplitudes of oscillation, respectively.

9.5 **Primary explosive propellant**

Primary explosive is the principal energetic ingredient of a solid propellant, because its critical diameter is sufficiently small and its burning rate is high to burn stably on a small scale. Requiring low brisance of primer explosive, lead styphnate (LTNR) and NHN are selected as two moderate energetic ingredients. The basic properties of the two explosives are shown in Table 9.1.

The primary explosive propellant is composed of an energetic compound (lead styphnate or nickel hydrazine nitrate), oxidant (AP), and energetic binder (nitrocellulose, NC). The primary explosive compounds are hazardous substances that detonate easily, but if the binder or oxidant is less sensitive, they become insensitive and difficult to initiate or detonate. The compositions of primary explosive propellants are listed in Table 9.2.

A 10 mm × 10 mm epoxy cube substrate drilled with a cylindrical hole between 0.7 and 2.0 mm diameter and 1.5 mm length is used as a simple thruster to evaluate the performance of LTNR and NHN primary explosive propellants. The thrust and the impulse are measured using the VHB device.

Table 9.1 Basic properties of lead styphnate (LTNR) and nickel hydrazine nitrate (NHN).

Property	LTNR		NHN	
Molecule structure				
Molecule formula	$C_6HN_3O_8Pb$		$[Ni(N_2H_4)_3](NO_3)_3$	
Oxygen balance	−0.188		−0.057	
Crystal density	3.02 g/cm^3		2.129 g/cm^3	[18]
Explosive heat	1910 J/g	[19]	1014 cal/g in 1 atm of air	[18]
			4238 J/g	[19]
Detonation velocity	4900 m/s at 2.6 g/cm^3	[18]	7000 m/s at 1.7 g/cm^3	[18]
	5.2 km/s			
Five-second explosion temperature	167°C	[18]	282°C	[18]
	265−280°C (DSC)	[19]	252−272°C	[19]

Table 9.2 Compositions of primary explosive propellants.

Propellant	Primary-E	AP oxidant	NC binder
LTNR-NC	90%−75%	0	10%−25%
NHN-NC	90%−75%	0	10%−25%
LTNR-AP	50%−30%	50%−70%	<2% (addition)
NHN-AP	80%−60%	20%−40%	<2% (addition)

Experimental results indicated that the thrust and impulse of LTNR-NC and NHN-NC propellants are positively proportional to the diameters of propellant grains, and the thrust (Fig. 9.9) and impulses (Fig. 9.10) of LTNR-NC propellants are higher than that of NHN-NC.

The comparison of thrust and impulse between LTNR/NC and NHN/NC are listed in Table 9.3. The thrust and impulse of LTNR/NC are about 5 times of NHN/NC when using the same diameter of propellant grain, indicating LTNR/NC is a better propellant.

AP is an important oxidant in solid propellant. LTNR and NHN are fuel-rich (or negative oxygen balance) energetic compounds, therefore the addition of AP can enhance the oxygen level of LTNR and NHN propellants. The experimental results of thrust (Fig. 9.11) and the impulse (Fig. 9.12) of LTNR/AP and NHN/AP propellants.

The thrust and the impulse of LTNR/AP and NHN/AP propellants are barely affected by the diameter of propellant, but they are affected by the primary explosive (LTNR or NHN) ratio to AP. The dependence of thrust and impulse on the content of AP and primary explosives are listed in Table 9.4.

The 50/50 ratio of LTNR/AP and the 60/40 ratio of NHN/AP are identified as the best ratios in achieving high thrust and impulse. Additionally, it is found that the thrust of LTNR/AP is higher than that of NHN/AP, while the opposite trend is observed for the impulse.

Comparing the thrust and the impulse of LTNR/AP with other primary explosive propellants (seen Tables 9.3 and 9.4), the best performing propellant is LTNR/AP with the content ratio of 50/50, in which its thrust and the impulse are 34.65 mN and 183.98 μNs, respectively.

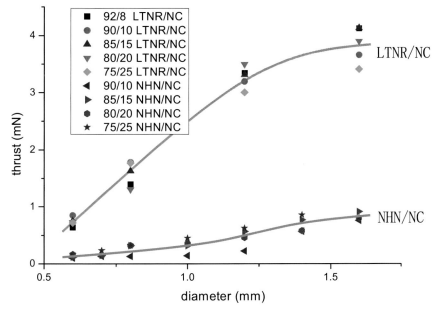

FIGURE 9.9

Thrusts of LTNR/NC and NHN/NC propellants in 0.6–1.8 mm diameters.

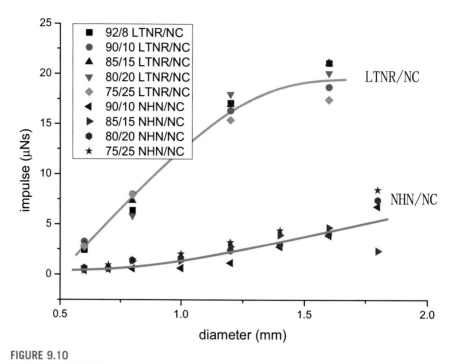

FIGURE 9.10

Impulses of LTNR/NC and NHN/NC propellants in 0.6–1.8 mm diameters.

Table 9.3 Thrusts and impulses of LTNR/NC and NHN/NC propellants with diameter.

Item	LTNR/NC				NHN/NC			
Diameter (mm)	0.6	0.8	1.2	1.6	0.6	0.8	1.2	1.6
Thrust (mN)	0.74	1.58	3.27	3.84	0.14	0.28	0.47	0.82
Impulse (μNs)	2.82	7.09	16.75	19.68	0.54	1.19	2.38	4.20

9.6 Nanothermite propellant

The thermite propellant can produce a small amount of high-temperature gas products, with most of them being vapor metal oxide or metal, such as Al_2O_3, resulting in a high molecular mass in the gas products. Some ingredients that produce gaseous hydrocarbon products are required to decrease the molecular mass, in which energetic binder and energetic compounds (NC, RDX, and HMX) are often selected as additives of thermite propellant.

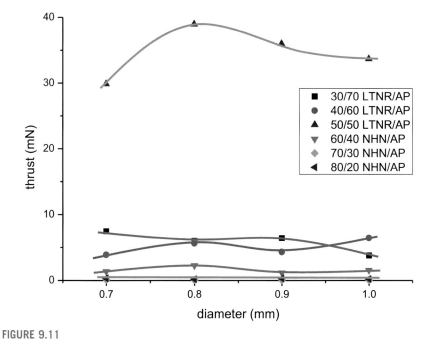

FIGURE 9.11

Thrusts of LTNR/NC and NHN/NC propellants in 0.7—1.0 mm diameters.

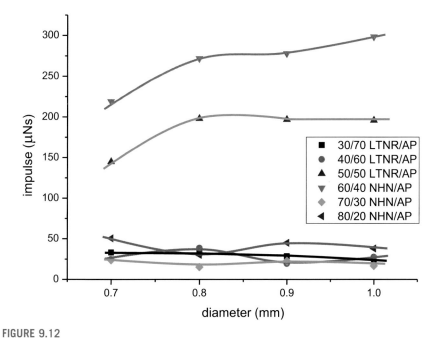

FIGURE 9.12

Impulse of LTNR/NC and NHN/NC propellants in 0.6—1.8 mm diameters.

Table 9.4 Thrusts and impulses of LTNR/AP and NHN/AP propellants with content ratio.

Item	LTNR/AP			NHN/AP		
Content ratio	30/70	40/60	50/50	60/40	70/30	80/20
Thrust (mN)	5.92	5.09	34.65	1.61	0.10	0.28
Impulse (μNs)	28.76	27.63	183.98	266.90	19.64	41.00

Thermite is a metal/metal oxide composite that releases intense heat and produces solid products when it is ignited. The suitable thermites for nanoenergetic propellant are Al/CuO, Al/MoO$_3$, Al/Fe$_2$O$_3$, and other potential thermites in stoichiometric proportion. Their properties are summarized in Table 9.4. Because of the small combustion chamber of microthruster, the volume heat of reaction per volume (cal/cm^3) is more important than the mass heat of reaction (cal/g).

The thermite performance can be calculated by the principle of free energy minimization and internal ballistic equations. The theoretical specific impulse of thermites listed in Table 9.5 are shown in Fig. 9.13. $I_{sp}(Optimum)$ is the theoretical specific impulse under standard conditions ($P_c/P_e = 70$, $P_a = 101\ kPa$, and standard rocket motor), and $I_{sp}(A_e/A_t = 1.0)$ is the theoretical specific impulse of straight nozzle motor.

The straight nozzle is not efficient, but it is the conventional structure for microthrusters as the nozzle length and chamber pressure of microthrusters is not sufficient to accelerate gas products to supersonic velocity, which is usually achievable in the efficient convergent-divergent nozzle of the solid rocket motor. The theoretical specific impulse of thermite is lower than 1000 m/s (or 100 s).

Table 9.5 Thermodynamic properties of high energetic thermites [20].

Reactants		Adiabatic reaction temperature	Gas production	Gas molecular mass	Heat of reaction	
Constitutes	ρ_{TMD} (g/cm^3)	Phase change (K)	Reactant gas (g)	Mole	cal/g	cal/cm^3
2Al+3CuO	5.109	2843	0.2431	78.42	974.1	4976
2Al + Fe$_2$O$_3$	4.175	3135	0.0784	96.15	945.4	3947
10Al+3I$_2$O$_5$	4.119	>3253	1.000	121.40	1486	6122
2Al + MoO$_3$	3.808	3253	0.2473	165.90	1124	4279
4Al+3PbO$_2$	7.085	3253	0.9296	363.40	731.9	5185
2Al + WO$_3$	5.458	3253	0.1463	491.08	696.4	3801

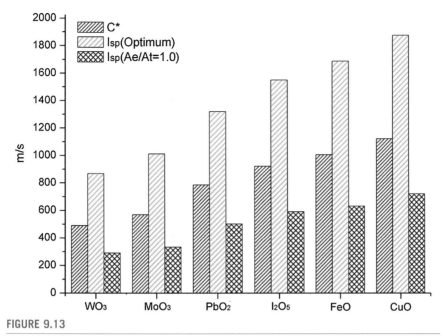

FIGURE 9.13

Theoretical specific impulses and characteristic velocity of thermites.

The specific impulse of Al/CuO is the highest among the thermites listed in Table 9.4, but its specific impulse is only 721.4 m/s (or 72 s) at $\varnothing = 1$.

The equivalence ratio is defined as Eq. (9.34).

$$\varnothing = \frac{(m_F/m_O)_a}{(m_F/m_O)_s} \tag{9.34}$$

where $(m_F/m_O)_s$ is the mass ratio of fuel to oxidant in the stoichiometric reaction requirement, $(m_F/m_O)_a$ is the mass ratio of fuel to oxidant in any constituents.

If $\varnothing < 1$, the constituent of thermite is in positive oxygen balance (oxidant-rich); if $\varnothing = 1$, it is in zero oxygen balance (in stoichiometry); if $\varnothing > 1$, it is in negative oxygen balance (fuel-rich).

The stoichiometric reaction releases the heat of the reaction maximally and its product temperature increases to the highest temperature. The formula of stoichiometry reaction of Al/CuO is

$$2Al + 3CuO \rightarrow Al_2O_3(l) + 3Cu(l, g)$$

If $\varnothing > 1$, there is not sufficient oxygen for aluminum to react fully, so some reactions happen in anoxic condition.

$$4CuO \rightarrow 2Cu_2O + O_2$$

$$2Al + 3Cu_2O \rightarrow Al_2O_3(l) + 6Cu(l, g)$$

The theoretical specific impulses of Al/CuO at different ratios ($\varnothing$) are shown in Fig. 9.14. The ratio of optimized performances of Al/CuO thermite is at $\varnothing = 0.9$, but the design principle of thermite should be fulfilling the needs of full reaction, or zero oxygen balance ($\varnothing = 1$).

The selection principle of thermite constitutes for the energetic propellant is high volume heat of reaction and long storage life. Al/CuO is a good thermite candidate, as it is a stable composite with high volume heat.

Although Al/CuO thermite is an energetic composite, it produces less gas species, and thus is not suitable as a propellant. Some energetic additives shall be mixed into Al/CuO thermite to produce extra gas for propulsion application. Some suitable additives are energetic binders (nitrocellulose (NC), polyvinylidene fluoride (PVDF)) and energetic compounds (RDX, HMX). The properties of NC, PVDF, RDX, and HMX are shown in Table 9.6.

The effects of these energetic additives on the theoretical specific impulse of Al/CuO ($\varnothing = 1$) are calculated by the minimum free energy method in the standard condition and straight nozzle ($A_t / A_e = 1$). The calculated results are shown in Fig. 9.15. The specific impulse shows a progressive trend with the content of energetic additives, but when the content of PVDF increases to and above 7.5%, the trend reverses. The effects of RDX and HMX are similar and better than NC. When the content of four additives is less than 5%, their effects are almost identical, giving a specific impulse of about 750 cm/s or 75 s.

Homogeneous mixing of nano-Al (nAl), nano-CuO (nCuO), and additives is a challenging but important process. Nano Al/CuO thermite is usually prepared by

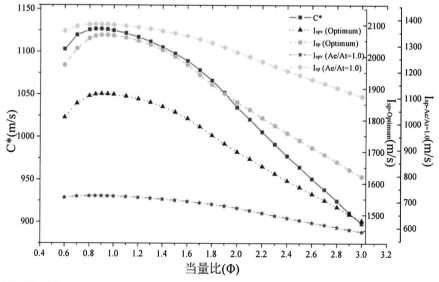

FIGURE 9.14

Theoretical specific impulses of Al/CuO at different fuel/oxidant ratios.

Table 9.6 Thermodynamic properties of NC, PVDF, RDX, and HMX.

Additive	Formula	Molecular mass	Standard heat of formation (kJ/mol)	Combustion heat (kJ/g)	Density (g/cm³)
NC [21]	$(C_6H_{7.583}O_{9.835}N_{2.418})_n$	270.9	-727.2	10.067	1.65
PVDF [22]	$(-CH_2-CF_2-)_n$	64.0	-375.62	14.900	1.77
RDX	$C_3H_6O_6N_6$	222.12	70.70	9.475	1.82
HMX	$C_4H_8O_8N_8$	296.16	75.02	9.334	1.90

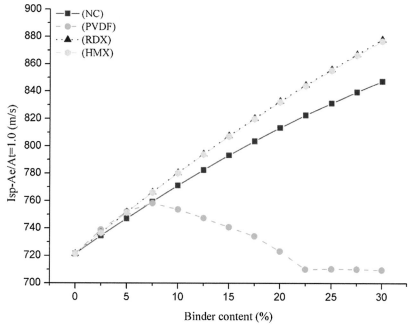

FIGURE 9.15

Theoretical specific impulse of Al/CuO composited additives.

ultrasonic mixing process (US), sol—gel process, molecular self-assembly process and electrostatic spraying (ES) process, and mechanical mixing (MM) process [23]. The technology map of preparing Al/CuO is shown in Fig. 9.16.

The sol—gel process is a standard synthesis process of nanoparticle CuO, the molecular self-assembly process is a good method to mix nAl and nCuO in a molecular scale, while ultrasonic mixing and electrostatic spraying are usually used to prepare multiingredient composites. Although the homogeneities of composites prepared by

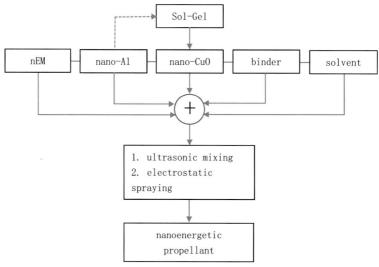

FIGURE 9.16

Preparation processes of thermite propellants.

the US and ES are lower than the molecular self-assembly process, the two processes are simple and easy to conduct to mix multicomponents of thermites. The sol–gel technology can also mix nAl with nCuO, in which nAl particles are mixed into the precursor of $Cu(NO_3)(OH)_3$ during the sol–gel process. The Al/CuO nanoparticles are synthesized after calcining the precursor of $nAl/Cu(NO_3)(OH)_3$ for 2 h at 300°C. The associated chemical reaction formula is

$$nAl + Cu(NO_3)(OH)_3 \rightarrow nCuO + nAl + NO_2 \uparrow + H_2O \uparrow$$

Al/Cu/NC thermite is a standard nanothermite propellant for microthruster. The morphologies of Al/Cu/NC composites prepared by ES process and MM were analyzed by electron microscope (EMS), as shown in Fig. 9.17. The dispersion of nAl and nCuO in the ES process is more homogeneous than that in the MM process. nAl and nCuO nanoparticles are well coated by nitrocellulose using the ES process.

The thermochemical properties of Al/CuO/NC were analyzed by differential scanning calorimetry (DSC) and are shown in Table 9.7. Comparing the reaction heats of Al/CuO/NC at the same content of NC, the samples prepared using ES demonstrated higher values than those prepared using MM by approximately 200 J/g. Al/CuO/NC-2.5% (ES) sample produces reaction heat rather similar to Al/CuO (MM). The decomposition temperature and exothermic temperature of samples from different methods and compositions are almost identical.

Comparing the morphologies of Al/CuO/NC and Al/CuO/PVDF prepared by electrostatic spraying process (Fig. 9.18), nanoparticles of nAl and nCuO disperse well in the NC film, but not in the PVDF film in which many nanoparticles aggregation was observed.

(a) Al/CuO(MM) ×10.0k

(b) Al/CuO(MM) ×50.0k

(c) Al/CuO/NC 10.0% (MM) ×10.0k

(d) Al/CuO/NC 10.0%(MM) ×50.0k

(e) Al/CuO/NC 10.0% (ES) ×10.0k

(f) Al/CuO/NC 10.0%(ES) ×50.0k

FIGURE 9.17

Morphologies of Al/Cu/NC composites prepared by electrostatic spraying (ES) process and mechanical mixing (MM).

A simple thruster, in the form of a semisealed combustion chamber of 1.00 mm diameter and 1.50 mm length, was fabricated using an epoxy substrate to evaluate the propulsion performances. Thermite powder was pressed into the cylindrical bore at 1.10 g/cm^3 or 1.3 mg.

Table 9.7 Thermochemical properties of Al/CuO/NC by electrostatic spraying process and MM processes.

Constitutes	Reaction heat (J/g)	Temperature of decomposition peak (K)	Temperature of exothermic peak (K)
Al/CuO (MM)	826.6	561.2	/
Al/CuO/NC-2.5% (ES)	857.3	561.8	195.8
Al/CuO/NC-5.0% (ES)	760.8	561.2	205.8
Al/CuO/NC-7.5% (ES)	732.0	561.8	207.8
Al/CuO/NC-10.0% (ES)	600.8	570.5	208.5
Al/CuO/NC-2.5% (MM)	680.5	562.5	199.2
Al/CuO/NC-5.0% (MM)	550.1	564.5	204.5
Al/CuO/NC-7.5% (MM)	438.2	569.2	208.7
Al/CuO/NC-10.0% (MM)	419.4	569.8	209.0

(a)Al/CuO/PVDF-5%(ES) ×10.0 k

(b)Al/CuO/PVDF-5%(ES) ×30.0 k

(c)Al/CuO/NC-5%(ES) ×10.0 k

(d)Al/CuO/NC-5%(ES) ×30.0 k

FIGURE 9.18

Morphologies of Al/CuO/NC and Al/CuO/PVDF prepared by electrostatic spraying process.

The flames of Al/CuO, Al/CuO/NC, and Al/CuO/PVDF ignited by laser are shown in Fig. 9.19. The flame of Al/CuO is a blasting flame that indicates an unstable burning. In contrast, the flames of Al/CuO/NC and Al/CuO/PVDF are smooth, with a brighter flame observed in Al/CuO/NC.

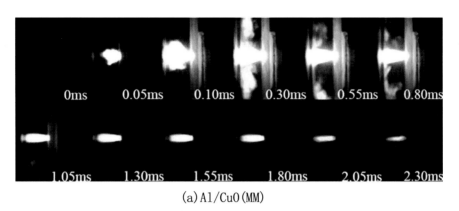

(a) Al/CuO (MM)

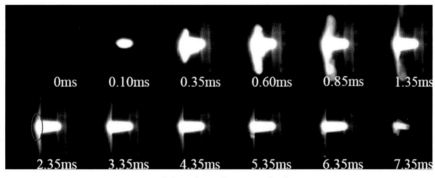

(b) Al/CuO/NC2.5% (ES)

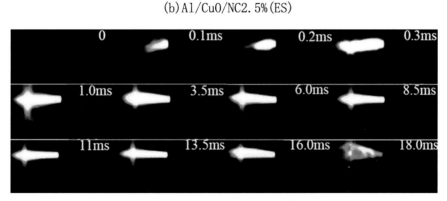

(c) Al/CuO/PVDF5% (ES)

FIGURE 9.19

Combustion flame of Al/CuO, Al/CuO/NC, and Al/CuO/PVDF.

The performance of specific impulses and combustion duration of thermite propellants were tested by the horizontal thrust balance. The performance of Al/CuO/NC prepared by electrostatic spraying process and mechanical mixed process is shown in Fig. 9.20 and Al/CuO/PVDF prepared by electrostatic spraying process in Fig. 9.21.

The mixing method affects the combustion performance of thermites. The combustion duration of Al/CuO/NC prepared by electrostatic spraying process is shorter than by mechanical mixed process but the specific impulse is higher, seen Fig. 9.20. For example, for 2.5% NC, the duration and the specific impulse of Al/CuO/NC prepared by the mechanical mixed process are 14.0 ms and 86.7 m/s, in comparison to 7.5 ms and 250.2 m/s as prepared by the electrostatic spraying process. This is mainly due to the efficient reaction in the well-coated nAl and. In preparing homogeneous thermite, the electrostatic spraying process is better than the mechanical mixing process.

The content of PVDF studied is 5%−25% (wt%). The combustion duration decreases while the specific impulse increases when the amount of PVDF is increased. The specific impulse increases to a maximum value of 77.1 m/s that is only 10.2% of the theoretic value (753.8 m/s). When the content of PVDF increases to 25%, the specific impulse decreases to below 34.2 m/s that is lower than 231.7 m/s of binder-free composite, suggesting the content of PVDF should not exceed 15%.

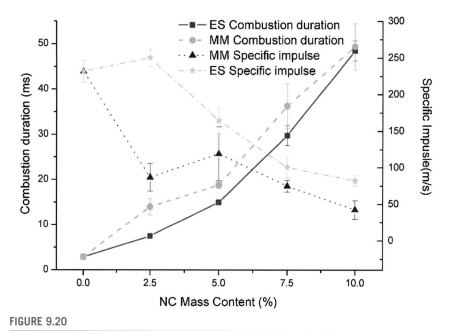

FIGURE 9.20

Performance of Al/CuO/NC at different content of NC.

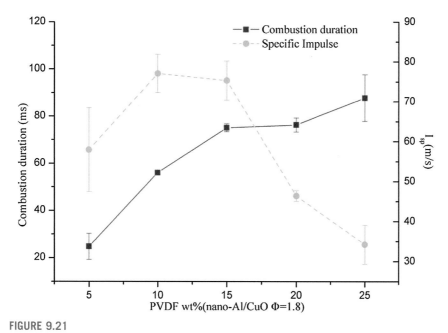

FIGURE 9.21

Performance of Al/CuO/PVDF at different content of PVDF.

When NC and PVDF binder are composited with Al/CuO, the combustion durations have prolonged to 48.5 ms (NC) and 87.8 ms (PVDF) from 2.9 ms as measured in the case of Al/CuO only. This is an attribute to the coating of binder on the nanoparticles of fuel and oxidant, which slows down the reaction rate.

9.7 Conclusion

Small critical diameter and insensitivity are required for the nanoenergetic propellants in solid chemical microthrusters. Nanotechnology is a useful way to enhance the burning stability of micropropellant grain in MEMS thruster with microscale chamber. Nanostructured energetic materials are promising for future use in MEMS thruster, but some traditional technologies are also viable, such as nanoprimary explosive propellant and nanothermite propellant. Nitrocellulose is a good energetic binder for the nanoenergetic propellants to decrease sensitivities of the propellants and to produce more gaseous products, but the content should be controlled over 15%. LTNR and NHN are possible nanoenergetic materials, with LTNR being superior in terms of output thrust and impulse. Ammonium perchlorate is a useful additive to enhance the impulse of propellant. The optimum content of AP in the primary explosive propellant is about 50%. Thermites are high heat energetic materials, but they cannot produce a lot of gas. Other additives of gas generators are required in the nanothermite propellants. The stoichiometric content of thermite releases the largest amount of heat, which is important for stable burning in small-scale propellant grain. Al/CuO is usually selected since it can release more heat than the other thermites. NC and PVDF can be used as a binder, with NC performing better than PVDF in terms of impulse. The burning duration of Al/CuO/Binder increases proportionally with the content of the binder, but the impulse decreases. To get a high impulse, the content of NC and PVDF binders should be below 5% and 15%, respectively.

Acknowledgments

We thank Dr. Kean-How Cheah for giving some good suggestions on the manuscript and helping to revise the manuscript of Chapter 9. We also thank Ms. Sha Long for contributing her time and carefully working to edit the manuscript and draw graphs.

References

[1] Shen, R., Wu, L., Qin, Z., Wang, X., He, N., 2017. New Concept of laser-augmented chemical propulsion. In: DeLuca, L., Shimada, T., Sinditskii, V., Calabro, M. (Eds.), Chemical Rocket Propulsion a Comprehensive Survey of Energetic Materials. Springer, pp. 689–696. https://doi.org/10.1007/978-3-319-27748-6.

[2] Zhang, H., Duan, B., Wu, L., et al., 2021. Actualization of an efficient throttleable laser propulsion mode. Energy 221, 119870.

[3] Barkley, S., Kindem, D., Zhu, K., et al., January 7–11, 2019. Dynamic control of composite solid propellant flame spread through microwave eddy current heating of propellant-embedded antennas. In: AIAA Scitech 2019 Forum. https://doi.org/10.2514/6.2019-1239. San Diego, California.

[4] Lynch, J., Ballestro, M., Barkley, S.J., Cazin, R., Michael, J.B., Sippel, T.R., January 5, 2016. Microwave-supported Plasma Combustion Enhancement of Composite Solid Propellants Using Alkali Metal Dopants. AIAA Aerospace Sciences Meeting, San Diego CA.

[5] Bao, L., Wang, H., Zheng, T., Chen, S., Zhang, W., et al., 2020. Exploring the influences of conductive graphite on hydroxylammonium nitrate (HAN)-based electrically controlled solid propellant. Propellants, Explos. Pyrotech. 45, 1790–1798.

[6] Bao, L., Zhang, W., Chen, Y., Chen, S., Shen, R., Ye, Y., 2019. Thermal decomposition and conductivity characteristics of HAN-based electrically controlled solid propellant. Chin. J. Energetic Mater. 27, 743–748.

[7] Shen, R., Ye, Y., Zhu, P., Hu, Y., Wu, L., Qin, Z., 2016. Nanostructured energetic materials and energetic chips. In: Vladimir, E.Z.A.R.K.O., Alexander, A.G.R.O.M.O.V. (Eds.), Energetic Nanomaterials Synthesis, Characterization, and Application. Elsevier Publications, pp. 139–161. ISBN: 978-0-12-802710-3.

[8] Shen, R., Ji, F., Wang, S., Zhang, W., Ye, Y., 2020. Reactivity and combustion of porous silicon energetic chips. Int. J. Energ. Mater. Chem. Propuls. 19 (2), 113–124.

[9] You, Z., Zhang, G., Ling, Y., 2005. Design and modeling of MEMS-based solid propellant propulsion. Opt Precis. Eng. 13 (2), 117~126.

[10] Rossi, C., Fabre, N., Conedera, V., Esteve, D., 1999. Design fabrication and thrust prediction of solid propellant microthrusters for space application. Proc. SPIE - Int. Soc. Optical Eng. 3680 (II), 906–916.

[11] Rossi, C., Esteve, D., Mingues, C., 1999. Pyrotechnic actuator: a new generation of Si integrated actuator. Sensor. Actuator. 74, 211–215.

[12] Lewis Jr., D.H., Janson, S.W., Cohen, R.B., Antonsson, E.K., 2000. Digital micropropulsion. Sensor. Actuator. 80, 143–154.

[13] Ma, L., Shen, R., Ye, Y., 2004. A study on the thruster properties of micro chemical thrusters charged with different propellants. Acta Armamentarii 25 (5), 547–550.

[14] Ye, Y., Shen, R., Xiao, G., Shan, Z., 2006. Study on the thrust test of micro-chemical thruster. Initiator Pyrotech. 1, 25–28.

[15] Wang, A., Wu, H., Tang, H., et al., 2013. Development and testing of a new thrust stand for micro-thrust measurement in vacuum conditions. Vacuum 91, 35–40.

[16] Koizumi, H., Komuraki, K., Arakawa, Y., 2004. Development of thrust stand for low impulse measurement from microthrusters. Rev. Sci. Instrum. 75, 3185–3189.

[17] Rocca, S., Nicolini, D., October 31 – November 4, 2005. Micro-thrust balance testing and characterization. In: The 29th International Electric Propulsion Conference, Princeton University. EPC, pp. 2005–2126.

[18] Zhu, S., Wu, Y., Zhang, W., Mu, J., 1997. Evaluation of a new primary explosive: nickel hydrazine nitrate (NHN) complex. Propellants, Explos. Pyrotech. 22, 317–320.

[19] Lao, Y., Sheng, D., 2011. The Science of Initiating Explosives and Relative Composition. Beijing Institute of Technology Press, Beijing, p. 231.

[20] Fischer, S.H., Grubelich, M.C., 1998. Theoretical Energy Release of Thermites, Intermetallics and Combustible Metals. SAND98- 1176C.

[21] Egan, G., Sullivan, K., Lagrange, T., et al., 2014. In situ imaging of ultra-fast loss of nanostructure in nanoparticle aggregates. J. Appl. Phys. 115 (8), 084903.

[22] Wang, H., Jian, G., Yan, S., et al., 2013. Electrospray formation of gelled nano-aluminum microspheres with superior reactivity. ACS Appl. Mater. Interfaces 5 (15), 6797–6801.

[23] Shen, R., Ye, Y., 2020. Special Effects and Applications of Energetic Materials. National Defense Industry Press, Beijing, pp. 39–83. ISBN 978-7-118-12226-8.

Solar sail as propellant-less micropropulsion

10

Yimeng Li, BEng [1], Kean How Cheah, PhD [2]

[1]*Graduate Student, School of Aerospace, University of Nottingham Ningbo China, Ningbo, Zhejiang, China;* [2]*Assistant Professor, School of Aerospace, Faculty of Science and Engineering, University of Nottingham Ningbo China, Ningbo, Zhejiang, China*

10.1 Historical background

10.1.1 Advantages and applications

Solar sailing is a propellant-less propulsion system that utilizes solar radiation pressure (SRP) and requires no expenditure of propellant [1]. The momentum is generated by reflecting solar photons from membrane sails which material is lightweight and highly reflective [2].

The interest in developing the solar sail for space propulsion applications increased over recent years as it is regarded as a key enabling technology for future space exploration missions or scientific demonstration tasks that are currently considered unaffordable in the budget when using chemical or electric propulsion systems [3,4]. Compared to the conventional chemical thruster, solar sails produce significantly less acceleration. However, they could conduct orbital plane changes more efficiently as they are propelled by an inexhaustible source of solar energy [1].

Since SRP acts continuously, solar sails have the potential to propel spacecraft to tremendous speed (even to heliocentric velocity), which allows for high maneuvers and long-duration exploration of the outer solar or even deep space system [2,5]. The spacecraft can be continuously accelerated by adjusting the orientation of the solar sail surface relative to the Sun, and they can hover indefinitely in space [6]. Another advantage of solar sails is the increase in the lifespan of the satellite as dependency on propellant for orbital maneuvers can be alleviated or eliminated. This makes some long-term missions that would initially not be feasible using the conventional thruster-based system more probable, such as near-Earth asteroid reconnaissance and sample return missions [1,5].

Solar sails can be used to achieve highly non-Keplerian orbits as they can perform station-keeping maneuvers that are not practical to conduct by traditional impulsive propulsion systems, such as heliocentric and planet-centered displaced orbits or artificial equilibria in the restricted three-body problem [7]. Solar sails provide a deorbit solution and strategy due to their low mass and require no

Space Micropropulsion for Nanosatellites. https://doi.org/10.1016/B978-0-12-819037-1.00008-6

273

propellant to maneuver, which have perspectives in aiding spacecraft attitude control and eliminating the impact of artificial defunct devices on operational satellites and space vehicles.

10.1.2 Historical development

It was in 1873 that James Clerk Maxwell predicted the existence of SRP, and Johannes Kepler also envisioned the possibility of sailing a spacecraft rigged with sails. Since then, different methods of capturing and utilizing the solar photon flow were proposed as possible solar sailing technology. However, the development of the solar sailing idea from theoretical studies into real engineering principles needs to be traced back to the late 1960s, when NASA launched an inflatable balloon satellite Echo II with a wide functional membrane surface to study the solar thrust effect.

The first in space solar sail concept was proposed by the Jet Propulsion Laboratory of NASA, which planned to construct a solar sailing spacecraft in the late 1970s, to rendezvous with Halley's Comet when it approached Earth closely in 1986. Although this program was not selected for funding and was canceled shortly, the conceptual study of a stabilized square sail in three-axis and solar electric propulsion led to the implementation of existing solar sailing technology [2,4].

Since the inception of solar sailing technology, it has undergone more than 50-year of theoretical studies and extensive laboratory explorations, before it was demonstrated in the space environment, Earth orbit, and on an interplanetary trajectory. In 1974, the solar sail concept was evaluated as a space propulsion method with Mariner 10 to Mercury planet. Although Mariner 10 was not designed for solar sailing, the orientation of solar panels was adjusted to utilize the propulsive force gaining from the photons in the same way of the solar sails principle [4].

Over the subsequent 20 years, numerous conceptual studies and analyses of dynamics and control strategies on solar sail were conducted. The Space Mirror Znamya 2, which deployment and structure shared similarities to that of a solar sail concept, was successfully developed by the Russian Space Agency. It was released from the Mir Space Station before launching from a resupply vehicle in 1993. Unfortunately, its second solar sail was shredded due to a collision with a spacecraft antenna 6 years later.

At the end of the 20th century, the German Aerospace Centre (DLR) in collaboration with the European Space Agency (ESA) and INVENT GmbH, constructed a ground testing (Fig. 10.1) [8] under weightlessness environment and ambient condition of a fully deployable light-weight solar sail of about 20 square meters on December 17, 1999. It demonstrated control and autonomous strategy and achieved a level 4 TRL [9].

More recently, three space agencies Japan Aerospace Exploration Agency (JAXA), the Planetary Society, and NASA developed different solar sail designs to test their propulsion capability in the Low Earth Orbit (LEO). Cloverleaf Sail was launched by JAXA into a 169-km orbit in 2004 to evaluate two methods of

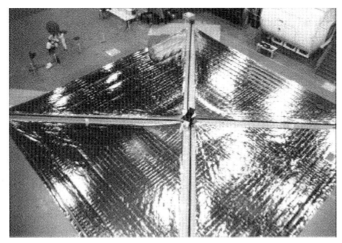

FIGURE 10.1

Ground testing of a fully deployable solar sail by DLR, ESA, and INVENT in 1999.

Reproduced with permission from M. Leipold, M. Eiden, C.E. Garner, L. Herbeck, D. Kassing, T. Niederstadt, T.
Krüger, G. Pagel, M. Rezazad, H. Rozemeijer, W. Seboldt, C. Schöppinger, C. Sickinger, W. Unckenbold, Solar
sail technology development and demonstration, Acta Astronaut. 52 (2) (2003) 317–326.

spinning deployment, while the generated propulsion was not clearly detected due to the atmospheric drag in this low orbit.

In 2005, Russian planned to launch COSMOS-1 that contained eight triangular sails assembled and funded by the Planetary Society and was considered as the first attempted solar sail spacecraft built for flight. However, its launch vehicle failed to reach the desired orbit at 800 km.

NASA renewed its study on solar sails in the early 20th century at the Glenn Research Centres in Ohio. In 2004, NASA funded the deployment and functional vacuum testing for two solar sailing prototypes both in the dimension of 20 square meters and was supported by two independently developed boom systems that utilized truss boom with a CP1 sail and inflatable boom with a Mylar sail [10]. However, technology development was stalled again as the proposed flight demonstration mission was not successfully selected in 2007. The essential information for several solar sail missions was summarized in Table 10.1.

10.2 Principle of operations

The photons impinging on the sail surface can be divided into absorbed photons, ρ_a, the specularly reflected photons, ρ_s, and the diffusely reflected photons, ρ_d. The ideal sail can be modeled as a flat surface that encounters both absorbed photons and reflected photons (ρ_s and ρ_d) [11]. The absorbed photons produce an impulse (a_{abs}) in the same direction as the incoming radiation (Fig. 10.2A) [1]. Based on the principle

Table 10.1 Summary of early solar sail missions.

Year	Mission	Space agency	Demonstrated techniques
1973	Mariner 10	NASA	SRP for attitude and orientation control
1976	Halley rendezvous (canceled)	NASA	Preliminary mission design
1993	Znamya 2	ROSCOSMOS	Spinning deployment and structural dynamics
1999	(Ground testing)	DLR, ESA, and INVENT GmbH	Control and autonomous strategy ground demonstration
2004	Cloverleaf	JAXA	Spinning deployment
2005	Cosmos 1 (fail)	Planetary society	Solar sail mission
2005	(Ground testing)	NASA	System-level solar sail ground demonstration

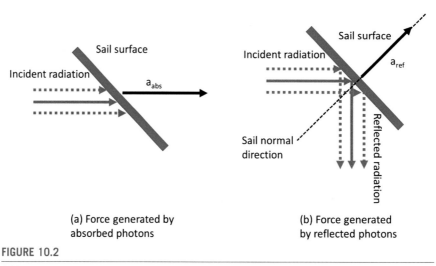

(a) Force generated by absorbed photons

(b) Force generated by reflected photons

FIGURE 10.2

SRP acceleration generated by absorbed and reflected photons.

of Newton's third law, the reflected photons ideally produce a momentum (a_{ref}) in the normal direction of the sail surface (Fig. 10.2B), which is a composition of the impulse produced by aborting the incident radiation and the impulse produced by emitting the reflected radiation [2].

The solar sail performance can be measured by the characteristic acceleration (a_c) and lightness number (β). a_c is the maximum SRP acceleration of solar sail at 1 astronomical unit (AU) distance from the Sun. β is the ratio of maximum SRP acceleration to the local gravitational acceleration of the Sun. A spacecraft under the influence of SRP force will acquire an acceleration level as determined by the areal density or sail loading (σ), which is the sail area to mass ratio. The relation between a_c (in mm/s^2), β and σ (in g/m^2) can be expressed as follows when assuming a perfectly reflective membrane [9]:

$$a_c = \frac{9.08}{\sigma} \quad \beta = \frac{1.53}{\sigma} \tag{10.1}$$

The pitch angle (α) is the angle defined between the normal sail surface ($\hat{n}$) and the sunlight direction ($\hat{s}$) [2]:

$$\hat{s} \cdot \hat{n} = \cos\alpha \tag{10.2}$$

The total acceleration can be obtained by dividing the SRP force by the total mass. For an ideally modeled spacecraft (the reflective efficiency parameter η is 1) that is perpendicular (pitch angle $\alpha = 0$) to the direction of sunlight, the SRP acceleration (a) can be calculated as follows [9]:

$$a = \frac{2PA}{m} = \beta \frac{\mu}{R_S^2} = a_c \left(\frac{R_E}{R_S}\right)^2 \tag{10.3}$$

where P is the SRP; A is the sail surface area; m is the total mass of solar sail; μ is the Sun's gravitational parameter; R_E is the distance between the Sun and the Earth; R_S is the distance between the Sun and the spacecraft.

The calculation results from Eq. (10.3) suggest an individual solar photon generates extremely small momentum and a solar sail can produce only about 9 N/km^2 of SRP force at 1 AU from the Sun even when its surface efficiency parameter η is 1 [9]. Considering the SRP force is the dominant force exerted on the solar sail, solar sail requires greater characteristic acceleration and lightness number than spacecraft using chemical or electrical propulsion. Thus, it can be concluded from Eq. (10.1) that a larger sail surface area and smaller total mass are crucial in design.

The total mass m can be divided into sail mass and hardware mass, m consists of the mass of sail and its support structure for deployment, bus, and subsystem for attitude control. A small sail could provide a characteristic acceleration of 0.1−0.4 mm/s^2 with a sail in the size of 15−30 m. While a membrane larger than 70 m^2 is required to achieve higher performance with a_c about 1 mm/s^2 [9]. It is worth to note that the a_c is also limited by σ, which means it cannot be inexhaustibly increased by expanding sail surface area. The theoretical limit of a_c is calculated to be 4.5 mm/s^2 for a typical σ of 2 g/m^2 [2].

10.2.1 Transfer trajectories

In the early study of a solar sail, analytical solutions were mostly applied to optimize the transfer trajectories due to the technical constraints of the numerical method and computational systems [12]. As computing technology has rapidly developed over the last decade, numerical techniques are widely employed, as it can obtain minimum-time solutions that are appropriate for several complex and specific missions, such as rendezvous, geostorm warning, and outer solar system missions [2].

There are two major differences in trajectory optimization between the solar sail and low-thrust electric propulsion. The solar sail generally used the minimum-time transfer trajectory rather than the fuel-optimal transfer trajectory. Additionally, the generated SRP force has constrained directions (opposite to the sunlight) while those of electric propulsion are free [5].

10.2.2 Solar sail non-Keplerian orbits

In non-Keplerian orbits, perturbations cannot be neglected. SPR forces generated by solar sail can be used for non-Keplerian orbits such as when the central planet is insufficiently nonspherical [13]. For example, the Sun-synchronous orbits are non-Keplerian orbits as the procession method of the orbits ascending node due to the nonspherical shape of the central body. In addition, to accomplish scientific objectives, seeking new orbits of solar sailing such as the artificial equilibria orbit for the geostorm mission and heliocentric displaced orbits for the polar observer mission is required [3].

10.2.3 Attitude control

Solar sail requires a subsystem for attitude control as it has asymmetric mass distribution, which means the moment of inertia needs to be balanced by active control torques [1]. The force direction can be controlled by adjusting the relative orientation of the sail surface to the sunlight, which can be achieved by changing the center of mass (c_m) or center of pressure (c_p) or utilizing a passive stability design.

The moment of inertia can be balanced by applying accessories or mechanisms and adjusting c_m or c_p to generate the required control torque. The objective of passive stability methods is to ensure sufficient restoring torques to maintain the attitude relative to the sunlight, which can be achieved by configuring the mass distribution or membrane shape [5].

10.2.4 Structural control

Slight changes in the sail structure could affect generated SRP thrust to a great extent as it depends on the sail surface shape. A large number of studies using analytical methods or numerical methods have been conducted to investigate the structural control in dynamic response during sail deployment [10]. The analytical method is widely applied since it has an advantage in solving complex structural design patterns and gaining more accurate responses. However, proposing analytical models for the square sail to conduct qualitative analysis is more challenging compared to the spinning solar sail [7]. Additionally, the finite element analysis method can be applied to generate solutions when the geometries or material properties of a sail are complicated.

10.3 Solar sail in CubeSat

Solar sails have advantages in providing continuous acceleration, longer lifetime, and generating non-Keplerian orbits, and they can also provide a deorbit solution. CubeSat-class solar sails have become a feasible way to advance more experiment and science missions at lower costs and wider operational lifespan [10]. A list of solar sail missions in CubeSat platform was summarized in Table 10.2, and elaborated subsequently.

10.3.1 NanoSail-D1

The NanoSail-D1 project was proposed by NASA's Ames Research Center in association with Marshall Space Flight Center (MSFC), which was responsible for the bus design and solar sail assembly, respectively. The team successfully developed a 1U CubeSat with 10 m^2 sail that aims to demonstrate the solar sail deployment and the utility for accelerating spacecraft deorbit in the LEO due to increased atmospheric drag [11].

After construction, it was then delivered to the SpaceX facility and launched by its Falcon 1 rocket in August 2008. Unfortunately, the Falcon failed to reach its orbit like the Cosmos 1 mission. The sail deployment mission failed, and the first NanoSail-D unit was lost.

10.3.2 NanoSail-D2

The first CubeSat to actually accomplish sail deployment in LEO is the NanoSail-D2 (Fig. 10.3), which was integrated into the MSFC's FASTSAT-HSV01 mission. The solar sail was stored aboard the FASTSAT, and it was expected to eject from the parent satellite on December 6, 2010. However, it did not successfully separate until

Table 10.2 Solar sail missions using CubeSat platform and their technical information.

Year	Mission	Space agency	CubeSat	Area (m^2)	Demonstrated techniques	Ref.
2008	NanoSail-D1 (failed)	NASA	1U	10	Sail deployment and deorbit	[10]
2010	NanoSail-D2	NASA	1U	10	Deorbit	[10]
2015	LightSail-1	Planetary society	3U	32	Propulsion	[13]
2019	LightSail-2	Planetary society	3U	32	Propulsion	[12]
2021	NEA Scout (planned)	NASA	6U	86	Propulsion	[10]

FIGURE 10.3

NanoSail-D2 with sail fully deployed.

Image credit: NASA.

January 19, 2011, due to an unsuccessful deployment commanded from the FAST-SAT. It was expected to demonstrate deorbit capability from the initial 650 km orbit and remain in LEO for 70−120 days before burning. However, it reentered the Earth after 243 days in orbit without active controls and then deorbit in the fall of 2011.

Three days after the ejection, the sail was automatically and successfully deployed as designed. The sail deployment mechanism was successfully demonstrated, and a similar mechanism was used by several other groups including the LightSail-1 due to its simple, robust design [13].

The NanoSail-D2 project was considered successful, and the project demonstrated the possibility of implementing a local constellation for satellites. It showed the effectiveness of using drag propulsion as a deorbit method, and it was also utilized to test upper atmospheric research in combination with the other scientific instruments [4].

10.3.3 LightSail project

The LightSail-1 (Fig. 10.4A) is the solar sail developed by the Planetary Society, and its objective is to provide a demonstration of sail deployment methods and system functionality. It was launched on May 20, 2015, and in orbit for 5 weeks. This project has advanced the TRL of the relevant subsystems to 6 and achieved all mission aims [9].

Another successful mission from the same space agency is LightSail-2 (Fig. 10.4B), which used a 3U CubeSat (Fig. 10.5) with a deployed area of 32 m^2 and aimed to demonstrate solar sail propulsion and precise attitude control by utilizing solar pressure such as using a momentum wheel to control the sail

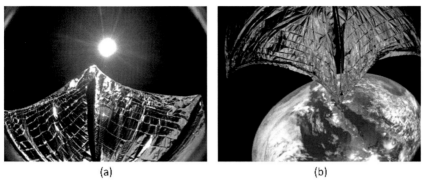

FIGURE 10.4

(A) LightSail-1 and (B) LightSail-2 with sail fully deployed.

Reproduced with permission from D.A. Spencer, B. Betts, J.M. Bellardo, A. Diaz, B. Plante, J.R. Mansell, The LightSail 2 solar sailing technology demonstration, Adv. Space Res. 67(9) (2021) 2878–2889.

FIGURE 10.5

LightSail-2, a 3U CubeSat with solar sail.

Reproduced with permission from D.A. Spencer, B. Betts, J.M. Bellardo, A. Diaz, B. Plante, J.R. Mansell, The LightSail 2 solar sailing technology demonstration, Adv. Space Res. 67(9) (2021) 2878–2889.

orientation [14]. Its launch was on-board the Falcon Heavy 9H, and it was released into a 720-km orbit where the effect of atmospheric drag is less significant [14]. It then raised about 600 m per day and deployed its sail [5].

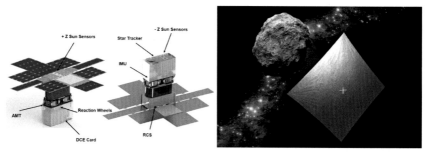

FIGURE 10.6

6U CubeSat NEA Scout with solar sail.

Reproduced with permission from J. Pezent, R. Sood, A. Heaton, High-fidelity contingency trajectory design and analysis for NASA's near-earth asteroid (NEA) Scout solar sail mission, Acta Astronaut. 159 (2019) 385–396.

10.3.4 Under development project

At present, the Jet Propulsion Laboratory of NASA in associate with MSFC aims to address Strategic Knowledge Gaps (SKGs) about information such as dimension, debris, orbit, and material properties of a 100 m near-Earth asteroid [10], which will be achieved by a 6U low-cost CubeSat solar sail (Fig. 10.6) [15]. In this Near-Earth Asteroid Scout (NEA Scout) mission, four 7-m booms and an aluminized polymer sail in 86 square meters will be utilized [3]. The NEA-Scout is planned for launch on the Space Launch System Exploration Mission 1 on Nov 01, 2021, and then deploy its solar sail within 3–4 days. This innovative project has the opportunity to become the first CubeSat to reach an asteroid.

10.4 Challenges and future

The performance comparison results between solar sail and typical spacecraft depend on the predicted technology level of a solar sail. Although several experiments have been conducted and a few CubeSat missions have been launched successfully, such as the NanoSail-D2 and Lightsail1 projects, their objectives are more focused on concept or subsystem demonstrating rather than functional validation of the entire system [14]. These missions have relatively low performance with a small acceleration generated. There are several key technologies that are crucial for future solar sail design and are also the focus of current research, including orbit dynamics and material technology [3].

10.4.1 Orbital dynamics

The accuracy of high-fidelity simulations accompanied by ground or on-orbit experiments are crucial in solar sail designs [6]. Many scholars have widely investigated the transfer trajectory using indirect methods to solve it from the Earth to the planets [5]. At present, the problem of optimizing the transfer trajectory of a single arc has been basically solved.

However, there are challenges in optimizing multiple-objective rendezvous using numerical methods [1]. When modeling the sail surface, most studies neglected the absorption and diffusion of photons, only perfect reflection was considered for the ideal model [4]. Additionally, the corrosive environment in space could affect the optical behavior of membrane sail, while this undesirable phenomenon is seldom considered in the trajectory design [16].

The future trend of orbital dynamic design in planning attitude control actuation could be directly using orbit-attitude coupling dynamics rather than attitude angles. This is because the calculation time will be significantly increased when solving multiscale problems in the coupling equation [16].

10.4.2 Material technologies

To minimize the sail loading and increase the performance level for the whole system, the material used in thin polymeric membrane and supported boom should be specifically selected and manufactured [7]. Currently, one side of the membrane sail is coated with highly reflective material for thrust generation and the other side is generally made of heat dissipation material [9]. The degradation effects due to the corrosion environment in space should also be taken into consideration.

The currently used material for the membranes are polymers, such as CP1 and ISASTPI. The areal density of the typical material is about $7 \ g/m^2$, and it ranges from 2 to $20 \ g/m^2$ [3]. New concepts of solar sail are required to achieve larger SRP acceleration. Removing polymer parts and using only thin metal membrane panels of thin metal films such as aluminum, magnesium, and beryllium is one of the proposed concepts [13]. Additionally, a new carbon fiber material was proposed by the Energy Science Laboratories, which claimed that it provides the same areal density as conventional material while is 200 times thicker [6].

However, there are few opportunities to test novel solar sail material designs, and the cost and risk of demonstrating these new material concepts are relatively high. According to reports, even the budget for low-cost missions such as LightSail-1 exceeds 1 million US dollars [11]. A new potential method that can reduce the risk of on-orbit verification of these new material concepts is the use of membrane spacecraft, which consists of thin-film devices using polymeric material with all components integrated as a whole surface [7]. Therefore, each CubeSat can hold hundreds of devices and a single low-cost CubeSat mission into dozens or even hundreds of such spacecraft, which can spread risks and reduce the cost of demonstration missions.

References

[1] Farrés, A., Jorba, À., 2016. Dynamics, geometry and solar sails. Indagat. Math. 27 (5), 1245−1264.

[2] Johnson, L., Sobey, A., Sykes, K., 2015. Solar Sail Propulsion for Interplanetary Cubesats.

[3] Hibbert, L.T., Jordaan, H.W., 2021. Considerations in the design and deployment of flexible booms for a solar sail. Adv. Space Res. 67 (9), 2716–2726.

[4] Fu, B., Sperber, E., Eke, F., 2016. Solar sail technology—a state of the art review. Prog. Aero. Sci. 86, 1–19.

[5] Huang, H., Zhou, J., 2019. Solar sailing CubeSat attitude control method with satellite as moving mass. Acta Astronaut. 159, 331–341.

[6] Caruso, A., Mengali, G., Quarta, A.A., Niccolai, L., 2021. Solar sail optimal control with solar irradiance fluctuations. Adv. Space Res. 67 (9), 2776–2783.

[7] Spencer, D.A., Johnson, L., Long, A.C., 2019. Solar sailing technology challenges. Aero. Sci. Technol. 93, 105276.

[8] Leipold, M., Eiden, M., Garner, C.E., Herbeck, L., Kassing, D., Niederstadt, T., Krüger, T., Pagel, G., Rezazad, M., Rozemeijer, H., Seboldt, W., Schöppinger, C., Sickinger, C., Unckenbold, W., 2003. Solar sail technology development and demonstration. Acta Astronaut. 52 (2), 317–326.

[9] Gong, S., Macdonald, M., 2019. Review on solar sail technology. Astrodynamics 3 (2), 93–125.

[10] Johnson, L., Young, R., Montgomery, E., Alhorn, D., 2011. Status of solar sail technology within NASA. Adv. Space Res. 48 (11), 1687–1694.

[11] Kennedy, R.G., Roy, K.I., Fields, D.E., 2013. Dyson dots: changing the solar constant to a variable with photovoltaic lightsails. Acta Astronaut. 82 (2), 225–237.

[12] Mansell, J., Spencer, D., Plante, B., Diaz, A., Fernandez, M., Bellardo, J., Betts, B., Nye, B., 2020. Orbit and Attitude Performance of the LightSail 2 Solar Sail Spacecraft.

[13] Berthet, M., Suzuki, K., 2021. Sensitivity of Solar Sailing Performance to Attitude-Orbit Interaction During the Lightsail 2 Mission.

[14] Spencer, D.A., Betts, B., Bellardo, J.M., Diaz, A., Plante, B., Mansell, J.R., 2021. The LightSail 2 solar sailing technology demonstration. Adv. Space Res. 67 (9), 2878–2889.

[15] Pezent, J., Sood, R., Heaton, A., 2019. High-fidelity contingency trajectory design and analysis for NASA's near-earth asteroid (NEA) Scout solar sail mission. Acta Astronaut. 159, 385–396.

[16] Rhatigan, J.L., Lan, W., 2020. Drag-enhancing deorbit devices for spacecraft self-disposal: a review of progress and opportunities. J. Space Saf. Eng. 7 (3), 340–344.

Hydroxylammonium nitrate—the next generation green propellant

11

Wai Siong Chai, PhD [1], Kai Seng Koh, PhD [2], Kean How Cheah, PhD [3]

[1]*Post-doctoral Researcher, School of Mechanical Engineering and Automation, Harbin Institute of Technology, Shenzhen, Guangdong, China;* [2]*Assistant Professor, School of Engineering and Physical Sciences, Heriot-Watt University Malaysia, Putrajaya, Malaysia;* [3]*Assistant Professor, School of Aerospace, Faculty of Science and Engineering, University of Nottingham Ningbo China, Ningbo, Zhejiang, China*

11.1 Historical development

An energetic material is defined as solid or liquid chemicals that provide stored chemical energy for various applications via proper triggering methods. Conventional energetic materials can be classified into four main categories including explosives, pyrotechnic compositions, propellants, and fuels/oxidizer mixtures [1]. Energetic materials are characterized by their sensitivity, also known as the ability to initiate a reaction from an energy stimulus as well as reactivity, that is, propagation and combustion characteristics after ignition. Ignition or initiation of an energetic material requires an energy stimulus that can be found in various forms such as impact, friction, electrostatic discharge or electrical spark, shock, and thermal heating. Using the conventional energy breakdown of chemical reactions involved, the total energy output of a particular energetic material can be precisely quantified. However, such energy level prediction varies with the combustion triggering method as a different reaction mechanism is involved with the varying method.

Rapid evolution works and research had been conducted since energetic material was first introduced in the 13th century as a black powder, which is known as gunpowder today for explosive purposes [2]. On the Earth, limited choices of energetic materials are available such as fossil fuel derived from hydrocarbon, synthesized alcohol, and explosives. These materials have a great impact on humanity's daily life especially in shortening the travel distance between countries in modern day's application. On the other hand, research and development on energetic materials had been greatly boosted by the blossom of the space industry since 1970s as intensive works in looking for alternative propellant options were attempted. As a result, various propellants such as solid propellants, liquid mono- and bipropellants, and hybrid propellants were proposed and tested [3,4]. However, energetic ionic liquids

Space Micropropulsion for Nanosatellites. **https://doi.org/10.1016/B978-0-12-819037-1.00005-0**

285

(EILs) previously studied in the late 1970s as liquid gun propellant (LGP) have been discarded for this peculiar application due to development problems [5].

Among the available options, monopropellants based on an aqueous solution of hydroxylammonium nitrate (HAN), with the chemical formula $[NH_3OH^+][NO_3]$, fall within the grouping of EILs and have received considerable attention due to their high energy density and green postcombustion characteristic. Furthermore, the HAN solution has properties such as high density, low viscosity, low freezing point, easier handling, and transport properties make it an attractive option as a replacement to conventional space propellants such as hydrazine and hydrogen peroxide [6]. Also, HAN solution exhibiting high solubility in aqueous media offers tuning properties as the solubility of HAN is as high as 95% in weight ratio, presenting in ionic form with cation NH_3OH^+ and anion NO_3^-. In addition, these green energetic EILs have advantages of low vapor pressures and reduced toxicities that make them a viable candidate for long-term storage and handling [7]. Although their long-term storage characteristics remain untested, particularly in comparison with the wealth of experience with hydrazine as a propellant, the EILs' promising physical properties combined with high energy densities and specific impulses make the development and evaluation of HAN-based propellants a relevant and rewarding arena of study [8]. Apart from its energy performance, HAN solution also demonstrated improved performance and operability, and improvement in safety and handling characteristics had led to increased interest in novel rocket monopropellants [8,9]. However, this solution is sensitive to both internal and external operational parameters such as metallic impurities, operating temperature, and pH changes that could potentially decrease the shelf life of the HAN solution [10].

The development of the HAN solution began in 1976 as a liquid gun propellant and was examined intensively in the 1980s for military applications. One of the best-known and familiar among other HAN formulations is LP1846, as seen in Table 11.1. It was first developed as an advanced liquid gun propellant with improved storability properties and reduced chemical aggressiveness and toxicity compared to hydrogen peroxide-hydrazine combinations [14,15]. In the early 2000s, the Air Force Research Laboratory has developed a liquid blend of HAN, HEHN, and water, known as AF-M315E, which offers 50% higher density-specific impulse (ρI_{sp}) than hydrazine [16] and has been used and tested as hydrazine substitute in the

Table 11.1 HAN-based monopropellant formulations with typical ingredients [11−13].

Propellant name	Ingredients (wt.%)		
LP1845	HAN (63.2)	TEAN (20)	H$_2$O (16.8)
LP1846	HAN (60.8)	TEAN (19.2)	H$_2$O (20)
HAN269MEO15	HAN (69.7)	MeOH (14.8)	AN (0.6) H$_2$O (14.9)
SHP163	HAN (74)	MeOH (16)	AN (4) H$_2$O (6)

Green Propellant Infusion Mission [17]. Besides, HAN-based monopropellant was also developed by Japanese researchers, namely SHP163, which consists of ammonium nitrate, methanol, and water [18] for space application.

11.2 Synthesis of hydroxylammonium nitrate

The interest in producing HAN to replace the hazardous rocket fuel, hydrazine, in the coming future has increased drastically. This is due to HAN having desirable characteristics and being labeled as a green alternative to conventional rocket fuel. The main synthesis routes available presently are acid—base titration, electrodialysis, and hydrolysis of oximes. The following section will describe in detail each synthesis method to understand the viability of using such methods, followed by an evaluation of each synthesis route of HAN that would be most beneficial with minimal economic and environmental impact.

11.2.1 Titration

One of the earliest reported synthesis routes for HAN is via titration between nitric acid (HNO_3) to a solution of hydroxylamine (NH_2OH) in a temperature-controlled environment (4°C). The method was widely adopted by many researchers due to its easy experimental setup. As the neutralization process during titration is highly exothermic, a low-temperature environment serves as a heat sink to maintain synthesis conditions by absorbing all heat release from titration between acid and base so that vaporization of pristine HAN is minimized (isothermal process) [19]. Upon completion of raw HAN production, the mixture produced (20 wt%) is evaporated using a rotary evaporator, to obtain a HAN solution at various concentrations (20—73 wt%). The chemical reaction and setup to prepare HAN through titration are shown in Eq. (11.1) and Fig. 11.1, respectively.

$$NH_2OH + HNO_3 \rightleftharpoons (NH_3OH)NO_3 \tag{11.1}$$

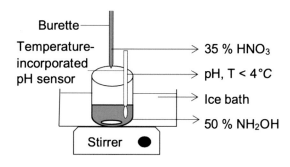

FIGURE 11.1

Schematic diagram setup for synthesis of HAN through titration [20].

11.2.2 Electrodialysis

Electrodialysis is the process whereby anions and cations in a solution move through a membrane in the presence of a potential difference. Positive-charge cations will be attracted to the anode and vice versa. The membrane then acts as a filter, to only allow anions to pass through that respective membrane when it is attracted to the cathode, and cations repelled and remain in the solution flow in that stage. The opposite is also applicable when the roles are reversed. Fig. 11.2 presents a visual representation of the process [21], in which the solution flow pathway can be observed, and the bars in between the pathways are the membranes. Wheelwright [22] first reported the electrodialysis process in stages that begins with feeding the salt, hydroxylamine sulfate ($(NH_3OH)_2SO_4$), in aqueous form, into a strongly acidic bed of nitric acid. This methodology is also used in the desalination of water and purification of plutonium [22].

11.2.3 Hydrolysis of oxime

In 2015, a new approach to synthesize HAN using hydrolysis of oximes was reported [23]. Using hydrolysis method, cyclohexanone oxime (CHO) is converted into HAN in the presence of an acid solution as catalyst.

This synthesis approach utilizes mixture of cyclohexanone oxime and dilute nitric acid that is mixed at a desired temperature (35°C) in a water bath in which the

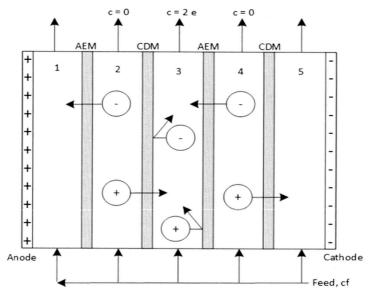

FIGURE 11.2

Visualization of the electrodialysis process when a potential difference is introduced within a membrane system.

chemical reactions will take place (Fig. 11.3). Upon completion of the reactions, the mixture will be extracted with ethyl acetate ($CH_3COOC_2H_5$), to remove excessive unreacted cyclohexanone and CHO, which will allow for the amount of HAN produced to be determined through redox titration with potassium permanganate.

11.2.4 Synthesis analysis

In terms of the yield of HAN from the synthesis methods, titration and hydrolysis of oximes provide high selectivity in producing HAN at 86.5% and 96.9%, respectively [23,24]. Both methods effectively produce high amounts of HAN and are fairly simple to conduct; however, there are some setbacks present for each method. Electrodialysis will not be considered in this evaluation, due to the limited research output in the literature to make a fair comparison. The advantages and disadvantages of each synthesis method are summarized in Table 11.2.

From the aforementioned methods of producing HAN, and summary of the advantages and disadvantages of each process, hydrolysis of oximes provides the greatest number of benefits with minimal drawbacks when compared to the remaining two methods. Reaction conditions such as pressure, temperature, and humidity, to name a few, are the main screening points for consideration when conducting the synthesis of HAN. Titration and the hydrolysis of oximes have reasonable temperature at which the reaction takes place at 4°C [20] and 35°C [23], respectively. Having such conditions will not put high dependence on external heating or cooling to produce HAN. However, attaining and maintaining the isothermal condition for the hydrolysis reaction will be more desirable, as it only requires minimal amount of heating from room temperature to 35°C, compared to cooling from room temperature to 4°C.

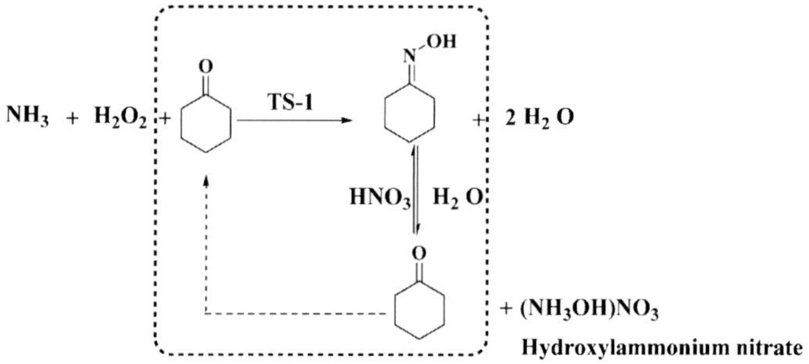

FIGURE 11.3

Overall process for the production of HAN using hydrolysis of cyclohexanone oxime in the presence acid catalyst.

Reproduced with permission from F. Zhao, K. You, R. Li, S. Tan, P. Liu, J. Wu, Q. Ai, H.A. Luo, Novel approach for the preparation of Hydroxylammonium Nitrate from the acid-catalyzed hydrolysis of cyclohexanone oxime, Ind. Eng. Chem. Res. 54 (3) (2015) 819–823.

Table 11.2 Summary of advantages and disadvantages of each synthesis method.

Method	Advantages	Disadvantages	Ref.
Titration	• Easy experimental setup • Increases selectivity from 60 to 65 wt% to 86.5 wt% • High N_2O content	• Spontaneous self-decomposition • Incomplete removal of catalyst • Ineffective scale up production	[19] [20] [24]
Electrodialysis	• Simple and efficient	• Risk of explosive reaction • Demanding reaction conditions	[21]
Hydrolysis of oximes	• Reasonable reaction conditions • Short reaction residence time • High conversion (62.9 wt %) and high selectivity (96.9%) • Cost effective for scale up production	• Reversible reaction in chemical kinetics • Solvents used for extraction are volatile and have low boiling point • Ratio of oxime to acid is limiting reactant in the reaction process	[23]

Both titration and hydrolysis selectively produce high amount of HAN. This is highly desirable, as these methods convert raw material to product effectively. However, in the case of titration, the catalyst being used in the process (commonly platinum) poses difficulties when attempting to filter the catalyst from the product, after the reaction is complete. Additionally, if a significant amount of catalyst is successfully removed, the binary mixture is then required to be neutralized with acid. As a result of neutralization, the possibility of spontaneous decomposition of the mixture can occur if temperature is not maintained at 4°C [20], which is undesired as the yield of HAN will reduce as a result. Shortcomings of hydrolysis revolve around the reaction being reversible (shown in Eq. 11.1). On the contrary, this opens an opportunity to produce more HAN (based on Fig. 11.3), as increasing the concentration of hydrogen ions present on the reactant side will favor the forward reaction, hence obtaining more HAN. Limitations in the extracting solvent used has nominal effects, as there are other solvents that give similar results of high HAN conversion and selectivity. Overall, economic viability of each method will be the driving force of upscaling the production. Under lab scale, hydrolysis is more cost effective [23] than titration, due to the points mentioned earlier.

11.3 Properties and safety evaluation

11.3.1 Physical properties, toxicity, and safety

HAN solution is known to be rich in charge as the charge density of the solution is associated with its solution concentration. The concentration is having direct relationship with its conductivity values [25], which is crucial in order for HAN to

Table 11.3 HAN molarity at different mass percentage.

Molarity (M)	Mass percentage (wt%)	Ref.
5	40	[26]
8.6	58	[27]
9.2	64	[27]
11	73.6	[28]
13	82	[29]

undergo combustion. Just like most common monovalent salt solution, HAN solution behaves with respect to the temperature and concentration dependence of the conductance. Although ion-pairing was probably minimal in aqueous HAN solution, the solubility of the solution can be increased when the solution is mixed with other solvent with lower dielectric constant, such as methanol. Higher HAN concentration yields higher conductivity values [25], with the specific conductance of HAN solution first increased with increased concentration and decreased after 7 M [25]. Lastly, the viscosity of HAN solution is also known to be increased with increment in concentration [25].

Density and molarity correlation is given as follows:

$$\rho_{HAN} = 1.12292 + 0.03099 \text{ M (8-15 M HAN)} \tag{11.2}$$

The molarity (M) and mass percentage (wt%) of HAN conversion and summary HAN physical properties are given in Tables 11.3 and 11.4, respectively.

LP1846 produced significant primary responses at concentration more than 25% for primary irritancy tests. Doses above 1 mg/mL are acutely toxic [32]. Short term, high concentration (14 days) and long-term, low concentration (13 weeks) toxicity tests of HAN in Wistar rats were studied. No adverse effects were observed during a 14-day period, while long-term exposure caused abnormal changes to tissue and the toxicity risk to mammals is not negligible [33].

Storage stability of HAN monopropellants for long service life on orbit is one of the most important problems due to the presence of metal ions such as Fe^{3+}. Material compatibility tests showed that some amount of metal ions elute from the surface

Table 11.4 Summary of HAN physical properties.

Physical property	Value	Ref.
Specific heat capacity for 1.63 M HAN	4.3 kJ/kg.K	[30]
Solubility in water	Highly soluble	[31]
Maximum molarity in solution	16 M	[31]
Conductance of 11 M HAN solution @ 24.7°C	0.155 S/cm	[14]
Conductance of 11 M HAN solution @ 53.8°C	0.250 S/cm	[14]

of fuel tank and tubes. The control of Fe^{3+} ion is required for propellant stability due to the possible redox reactions between HAN and metal ions. The contaminated HAN that was treated with the diethylene triamine pentaacetic acid chelating agent had similar performance as HAN, indicating good absorption of Fe^{3+} ions, which was confirmed by analytical methods. The best way to uptake Fe^{3+} ions from HAN solutions was the dilution of contaminated solutions through water, followed by the addition of chelating agent; then, the solution should be reconcentrated to desired final concentrations. The ability of iron species chelating agents was strongly dependent on the initial HAN concentration [34].

11.3.2 Vibration frequencies of HAN

Vibrational frequencies of solid HAN at 297 K were reported along with HAN-d, with observed frequencies assigned to each band. The hydrogen bonding of molten HAN was less than crystalline HAN [35]. The modes remaining at 470 K result from a small residue of NH_4NO_3 that has been identified before as a decomposition product of HAN [36]. Solid HAN exhibited more distinguished peaks in Raman analyses than HAN solution. In dilute HAN solutions, solvated ions were favored over contact-ion pairs [37], while ion-pairing became evident and eventually dominated the spectra above 8 M HAN [31].

The bands at 716, 830, 1050, and 1380 cm^{-1} were assigned to NO_3^- ion. The interaction between NH_3OH^+ with NO_3^-/H_2O resulting in NH_3^+ degenerate rock, NH_3^+ symmetric deformation, combination band, NH_3^+ symmetric stretch, and OH stretching near 1179, 1512, 2720, 2955, and 3150 cm^{-1}, with the last three bands shifted to higher frequencies with decreased HAN concentration. The infrared spectra at low HAN concentrations were dominated by a broad water band system at between 3200 and 3400 cm^{-1} [31]. Frequencies at 1655 cm^{-1} and 3000−3750 cm^{-1} were assigned to water component, with the latter being peaked at 3420 cm^{-1}. Both HA^+N-O symmetric stretch (N−OH vibration) at 1010 cm^{-1} and nitrate ion N−O symmetric stretch (v1 peak) at 1048 cm^{-1} were assigned to NH_3OH^+ and NO_3^-, respectively, and should be most useful in analytical work, as evidenced by the linearity above 8 M HAN solution [31,37].

Bands at 1004, 1044, and 1285 were assigned to N−OH, NO_3^- and N_2O, respectively [30]. Bands at 2230 and 2730 were assigned to N_2O and NH_3^+ bending combination, respectively [26]. Raman spectra of 11 M HAN solution were tabulated as below, with N_2O produced in decomposition products at 1290 and 2225 cm^{-1}, assigned to symmetric and asymmetric stretch mode, respectively [28]. The summary for the vibration frequency bands for HAN solution is provided in Table 11.5 and Fig. 11.4 [38].

Table 11.5 Wave number and respective assignments for HAN solution.

Wavenumber (cm^{-1})	Assignment
730	NO_3^- deformation
1010	N—OH stretch
1050	NO_3^- symmetric stretch
1190	OH bend (from HAN), NH_3 rock
1420	Asymmetric NO_3^- stretch
1620	NH_3 deformation
2730	NH_3, rock + deformation
3000	NH_3 stretch
3200	OH stretch (from both HAN and H_2O)
3230	OH stretch (from H_2O)
3600	OH stretch (from H_2O)

Typical infrared spectra of 80 wt% HAN solution are displayed as follows (Fig. 11.4):

11.3.3 Detonation and autocatalysis

Normal HAN decomposition results in nitrous oxide:nitrogen gas ($N_2O:N_2$) ratios of 2:1 to 4:1; however, in the presence of iron, this ratio has been determined to be 36:1 [23]. The presence of iron ion exhibited catalytic effect on HAN autocatalytic reaction (Liu et al., 2009) and the possible redox reactions between HAN and metal ions proceed as follows to produce nitrous oxide and nitrous acid (Eqs. 11.3 and 11.4) with the overall reaction displayed in (Eq 11.5) [39]:

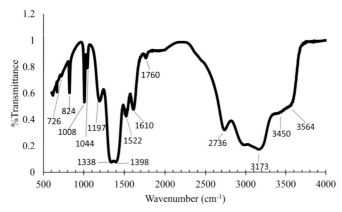

FIGURE 11.4

Infrared spectrum of 80 wt% HAN solution.

Reproduced with permission from W.S. Chai, K.H. Cheah, H. Meng, G. Li, Experimental and analytical study on electrolytic decomposition of HAN water solution using graphite electrodes, J. Mol. Liq. 293 (2019) 111496.

$$2 \, NH_3OH^+ + 4 \, Fe^{3+} => 4 \, Fe^{2+} + N_2O + H_2O + 6 \, H^+ \qquad (11.3)$$

$$2 \, Fe^{2+} + HNO_3 + 2 \, H^+ => 2 \, Fe^{3+} + HNO_2 + H_2O \qquad (11.4)$$

$$2 \, HAN => N_2O + 3 \, H_2O + 2 \, HNO_2 \qquad (11.5)$$

The autocatalytic decomposition of 24 wt% HAN showed two stages: a slow initiation stage and a fast propagating stage, which occurred very suddenly and quickly. The explosion stage can start at a much lower temperature using an aged sample. Stainless steel and titanium could accelerate the decomposition at lower temperatures compared to glass. HAN must be handled at low temperatures, and long-term storage should be avoided because of its autocatalytic decomposition behavior [40]. The autocatalytic decomposition of HAN was caused by HNO_2 [41−43], in which decreased induction time with increased HAN concentration supported a critical catalytic intermediate concentration developed before the rapid exothermic process commenced [30]. Strong autocatalytic behavior was exhibited in more concentrated HAN solution [44].

Fresh 24 wt% HAN solution without metal presence can be safely stored and used in room temperature. Heat of reaction of HAN is about 118 kJ/mol. Dissociation step was rapid above 175°C. Main process of HAN decomposition is NH_3O oxidation by HNO_3 and HNO_2 at the experimental conditions [45] and concluded that 187°C is critical temperature for storage of fresh diluted HAN solution [30,45].

The reactive intermediates such as HNO_2 could build up and result in performance degradation and autocatalysis [30], causing thermal runaway and leads to safety issues, which should be monitored in stored solutions of HAN. The overall autocatalytic reaction process is given below, showing the conversion of nitric to nitrous acid, which followed second-order kinetics at acid concentrations from 3 to 8 M [46]:

$$2 \, HNO_3 + NH_2OH -> 2HNO_2 + H_2O \qquad (11.6)$$

From Eq. (11.6), the HNO_2 dissociation step was found to be possessing lowest energy barrier of the rate controlling step and the was most favored dynamically. On the other hand, NH_2OH and HNO reaction had the highest energy barrier and was least favored dynamically [47].

On the other hand, it is a common fact that water content should contain more than 20% for the composition to be free from detonability. At the same time, high concentration presence of TEAN should be avoided because of the higher flame temperature and viscosity. It is therefore important to ensure the fuel-lean/oxidizer-rich composition is optimal [48]. Despite other HAN formulations such as SHP163 and AF-M315E showed no detonation risk, prolonged exposure to high temperature will initiate irreversible thermal decomposition over time [49,50].

11.4 Catalytic combustion of HAN

Catalytic combustion of HAN has shown to be promising in space propulsion applications. Various efforts have been dedicated in studying the suitable catalyst for effective decomposition of HAN-based propellant. Catalytic decomposition of HAN-based liquid propellant (LP1846) was first demonstrated with iridium-based catalyst (Shell 405 catalyst, 30% Ir/Al_2O_3) as it is a well-established and widely used catalyst in the space industry for hydrazine decomposition at temperature below 1100°C [8]. However, high combustion temperature of HAN (theoretical value of 2093°C) has shortened the iridium-based catalyst life due to sintering of catalyst [8,51]. Moreover, reaction rate was found to be lower than that of hydrazine as a result of lower iridium-induced catalytic activity on HAN propellant [52].

11.4.1 Reaction mechanism

For thermal as well as catalytic decomposition of HAN-based propellants, the difference in product gas species is not obvious. In both cases, the primary product gas species are nitrogen N_2 (major, thermodynamic product) and nitrogen oxide NO (medium, kinetic product) while secondary product gas species are nitrous oxide N_2O (medium) and nitrogen dioxide NO_2 (traces). HNO_3 is present as trapped solution. This suggests the possibility of similar reaction kinetics, which was subsequently validated as the reactions in catalyzed systems [53] were found to be similar to the thermal decomposition mechanism proposed by Lee and Litzinger [54], indicating that catalyst does not change the product gas species but only reduces onset temperature [55]. Catalyst was found to increase the preexponential factor of the early step in HAN decomposition reaction [56].

Differing from the two-step reaction in thermal decomposition, catalytic decomposition of HAN manifests a single-step reaction with higher reaction rate (Fig. 11.5) [57]. Thermal decomposition of HAN starts only when water has been entirely removed. However, the presence of an appropriate catalytic active phase leads to a drop of the onset temperature and triggers the decomposition at lower temperature, even in the presence of water [18]. Possible paths of thermodynamic and kinetic decomposition reactions involving catalyst are the following [57]:

$$NH_3OHNO_3 \ (aq) -> N_2 \ (g) + O_2 \ (g) + 2H_2O \ (l) \tag{11.7}$$

$$NH_3OHNO_3 \ (aq) -> NO_2 \ (g \ or \ aq) + 0.5N_2 \ (g) + 2H_2O \ (l) \tag{11.8}$$

$$NH_3OHNO_3 \ (aq) -> 2NO \ (g) + 2H_2O \ (l) \tag{11.9}$$

$$NH_3OHNO_3 \ (aq) -> N_2O \ (g) + 0.5O_2 \ (g) + 2H_2O \ (l) \tag{11.10}$$

Recent studies showed that the role of iridium catalyst is to increase the preexponential factor of the proton transfer reaction [56], which confirms the hypothesis that the HAN decomposition is limited by the proton transfer reaction [58]. In the

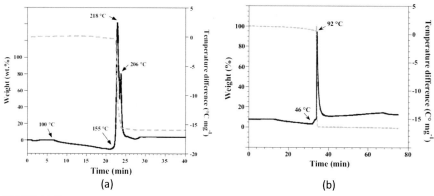

FIGURE 11.5

(A) Thermal and (B) catalytic decomposition of 95 wt% HAN solution.

Reproduced with permission from R. Amrousse, K. Hori, W. Fetimi, K. Farhat, HAN and ADN as liquid ionic monopropellants: thermal and catalytic decomposition processes, Appl. Catal. B Environ. 127 (2012) 121–128.

presence of Ir catalyst, NO formation is enhanced by H loss, as shown in the following equation [59]:

$$NH_2OH -> NO + 3/2\ H_2 \tag{11.11}$$

Another study indicated that ceria-based catalyst enhanced the HAN decomposition, as shown in the following equation [60]:

$$2\ Ce^{3+} + 3\ H^+ + HAN -> 2\ Ce^{4+} + NH_3OH^+ + HONO + H_2O \tag{11.12}$$

$$Ce^{4+} + NH_3OH^+ -> Ce^{3+} + 1/2\ N_2 + H_2O + 2H^+ \tag{11.13}$$

11.4.2 Development in catalyst

Identifying a suitable catalyst material to lower the onset temperature is crucial for catalytic decomposition of HAN-based propellant. Shell 405 catalyst (Fig. 11.6) is very active for HAN decomposition but deactivates after several firing tests due to high exhaust gas temperatures and oxidation medium [61–63]. Over 8000 s of cumulative operation have been demonstrated in single laboratory catalytic thruster using S405 catalyst, showing high restartability of HAN and catalyst [51]. However, adherence of active metal in Shell 405 catalyst was observed for a 30,000 s testing in a test chamber, with a rapid decline of catalytic activity after 20,000 s [52].

A systematic screening was done by simulating the adsorption of LP1846 propellant on a group of potential catalyst metals. Iridium was revealed as the most promising catalyst metal as it produced the best adsorption result among five noble metals (Ru, Rh, Pd, Ir, and Pt) [52]. Other than simulation, numerous experimental works

FIGURE 11.6

S405 catalyst (30% Ir/Al$_2$O$_3$).

Reproduced with permission from R. Amrousse, T. Katsumi, N. Azuma, K. Hori, Hydroxylammonium nitrate (HAN)-based green propellant as alternative energy resource for potential hydrazine substitution: from lab scale to pilot plant scale-up, Combust. Flame 176 (2017) 334–348.

were carried out to demonstrate the decomposition of HAN-based propellants using different noble metals as catalyst material. 80% HAN solutions were decomposed using platinum-doped alumina catalyst (Pt/Al$_2$O$_3$) at near room temperature (40°C), with short ignition delay (<1 s) [64]. Later, Ir-based catalysts were found effective in reducing onset temperature of HAN decomposition [55] as Ir/SiO$_2$ catalyst demonstrated room temperature ignition (20.7°C) [65]. Lower decomposition temperature and larger pressure slope were achieved using Ir-based catalyst (51°C, 260 mbar/s) [57] as compared to Pt-based catalyst (73°C, 134 mbar/s) [66], owing to the better activity and larger surface area (230 vs. 63 m^2/g) of Ir-based catalyst.

Recently, cerium oxide doped with cobalt (CeCo) was explored as an alternative catalyst material for HAN decomposition. The catalyst has demonstrated better durability in terms of physical integrity and resistant to catalyst poisoning than the existing Ir/Al$_2$O$_3$ catalyst. It was found that the presence of Ce^{3+} in ceria matrix is critical for its enhanced catalytic activity in HAN decomposition. The other benefits of high temperature tolerance and initiation of decomposition reaction with high exothermicity at low temperature (120–130°C) make it a promising catalyst [53].

A bimetallic catalyst of Ir–CuO was studied and demonstrated a rather similar performance to other mono Ir-based catalysts but offered the advantage of economically cheaper [18]. Another study utilized a bifunctional catalyst consisting of both Ir and CeCo (Fig. 11.7), displaying higher catalyst activity and longevity over individual catalyst [67].

Apart from using traditional catalyst where the catalyst materials are coated on the ceramic support, the HAN decomposition could be catalyzed by blending a small

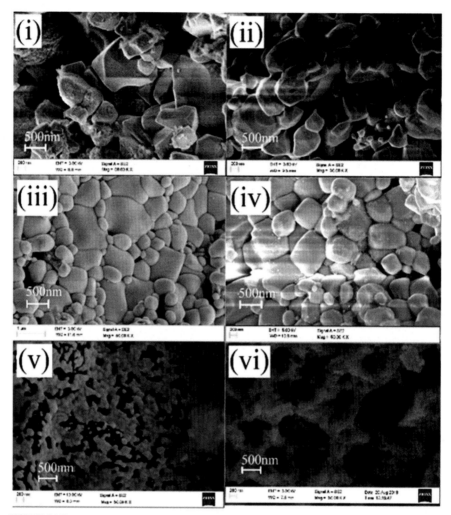

FIGURE 11.7

SEM images of final pellet form for different catalyst (i) CeCo 26, (ii) CeCo 26x, (iii) Ir/CeCo 26, (iv) Ir/CeCo 26x, (v) Ir/γ-Al$_2$O$_3$, and (vi) Ir/γ-Al$_2$O$_3$x.

Reproduced with permission from R. Agnihotri, C. Oommen, Evaluation of hydroxylammonium nitrate (HAN) decomposition using bifunctional catalyst for thruster application, Mol. Catal. 486, (2020) 110851.

amount of nanoparticles into the HAN-based propellant. Addition of carbonized rice husk-derived potassium hydroxide (CRH-KOH) activated carbon changed the combustion characteristics of 95 wt% HAN-water solution, where the double-stage thermal decomposition became single stage with reduction of decomposition temperature by 40–45°C. The burning rates were improved [68] as the high specific surface area of activated carbon, 3000 m^2/g, provides defective surfaces and edges for the formation and stabilization of free radicals (OH radicals), which are formed during the decomposition of nitric acid [54]:

$$HNO_3 -> NO_2 + OH \qquad (11.14)$$

In addition, a 30% reduction in NO_x gases was achieved with activated carbon addition when compared to the case of pure HAN solution. The onset temperature has been reduced from 185°C (HAN-iridium catalyst) to 86°C (HAN-CRH-KOH-activated carbon, indicating the possibility to reduce if not eliminate the use of noble metals as catalyst in the future [69]. Similar improvement in overall burning rate (500%−830%) of HAN propellant has been reported when SiO_2 and TiO_2 nanoparticles [70] were added at mass loading of 1%−3%. Nevertheless, it is noteworthy that such enhancement diminishes at higher pressures.

Powdered and shaped catalysts remained active after 23 injections at 45°C, displaying noninhibiting effect of shaping (Courthéoux et al., 2005). In an isothermal test at 50°C, powdered catalyst (50 μm) displayed best activity with a complete decomposition of HAN solution, by having good contact with monopropellant, as compared to sphere catalyst (1.5 mm). The presence of preferential paths on the catalyst bed causes the propellant to have less contact with catalyst, indicating that spherical-shaped catalyst is not suitable as catalyst bed [71].

High reactivity could be achieved with grain catalyst (130 m^2/g) compared to monolith catalyst (22 m^2/g) due to good contact between solid and liquid phases and enhanced heat and mass transfer [61]. Catalyst wear rate with respect to total flow rate for the propellant was 6% for fine particles and 2% for coarse particles. The catalyst experienced greater wear than the case with hydrazine [72].

Impregnated metal catalysts (2−5 nm, Fig. 11.8) exhibited higher catalytic activity than one-step addition (hundreds nm) as the size of metal crystallites is smaller in impregnation method [64]. However, another study showed that higher dispersion and smaller particle size (1.3 vs. 2.4 nm) are not favorable for obtaining high activity, indicating that the effect of particle size toward activity is not indefinite and the particle size might be optimum at 2 nm [65].

Onset temperature was found to be the lowest when employing catalyst with both support and active phase catalyst (75°C), as compared to the case of catalyst with only support (105°C) and without catalyst (128°C) [73]. Unsupported Pt group metals were proven to be the best metal catalyst among the injection testing of HAN solution in earlier years [52]. Onset decomposition temperature for Pt/Al_2O_3 catalyst was higher than 40°C [64].

Although aerogel-treated Al_2O_3 support has higher specific surface area, has better thermal stability at high temperature (1200°C), and is more homogeneously dispersed than xerogel-treated Al_2O_3 support, aerogel-treated Al_2O_3 support displayed lower catalytic activity at low temperature and higher onset temperature in comparison. This might be attributed to the catalytic surface activity of surface acidic centers [64]. This shows that different treatments for the same metal catalyst and support show different results, such as higher reaction rate or reduced decomposition temperature [66].

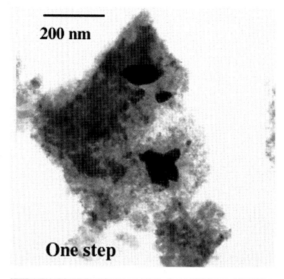

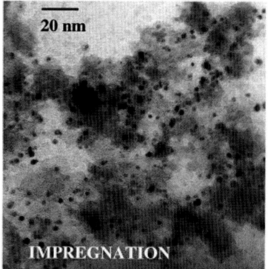

FIGURE 11.8

Transmission electron micrographs of xerogel catalysts. Comparison of the one-step procedure and the impregnation procedure.

Reproduced with permission from L. Courthéoux, F. Popa, E. Gautron, S. Rossignol, C. Kappenstein, Platinum supported on doped alumina catalysts for propulsion applications. Xerogels versus aerogels, J. Non-Cryst. Solids 350 (2004) 113–119.

11.5 Challenges and future perspectives

Room temperature catalytic decomposition of HAN could be achieved with Ir-based catalyst, but it did not complete in single step, with the remaining HAN solution to be completely decomposed at 80°C [65]. Thus, other key factors that influenced iridium catalytic activity need to be studied. Complete decomposition of HNO_3 is required to achieve best performance of HAN-based monopropellant [55]. Further work in reducing load on catalyst and improving catalyst charge method is required to reduce wear on catalyst [72]. Although CeCo catalyst demonstrated more resistant to catalyst deactivation than commercial Ir-based catalyst, its long-term decomposition stability and performance in microsatellite operation (1−2 years) under high combustion temperature were still unexplored. The future of HAN being used as a total replacement to the conventional propellant remains a challenge even though there has been many lab bench or small-scale application, particularly in space-related application being demonstrated. However, its current performance has shown promising result as a potential candidate of monopropellant. The future is bright for the continuous development of this propellant where the sky will be the limit for this propellant to be utilized in space missions.

References

[1] Millar, D.I.A., 2012. Energetic Materials at Extreme Conditions. Springer Berlin Heidelberg.

[2] Gray, E., Marsh, H., McLaren, M., 1982. A short history of gunpowder and the role of charcoal in its manufacture. J. Mater. Sci. 17, 3385−3400.

[3] Chou, S.K., Ang, S.S., 2004. MEMS-based solid propellant microthruster design, simulation, fabrication and testing. J. Microelectromech. Syst. 13, 165−175.

[4] Zwick, E.B., Calif, P., 1957. Reaction Chamber for Monopropellant and Bipropellant Fuels. US Patent 2999358.

[5] Klingenberg, G., Knapton, J.D., Morrison, W.F., Wren, G.P. (Eds.), 1998. Liquid Propellant Gun Technology, Progress in Astronautics and Aeronautics. American Institute of Aeronautics & Astronautics, Reston, Virginia.

[6] Kappenstein, C., Pillet, N., Melchoir, A., 2002. New nitrogen-based monopropellant (HAN,ADN,HNF,…). physical chemistry of concentrated ionic aqueous solutions. In: International Conference on Space Transportation for the XXI Century, pp. 16−125.

[7] Chowdhury, A., Thynell, S.T., 2010. Kinetics of decomposition of energetic ionic liquids. Propellants, Explos. Pyrotech. 35 (6), 572−581.

[8] Wucherer, E.J., Christofferson, S., Reed, B., 2000. Assessment of high performance HAN-monopropellants. In: 36th AIAA/ASME/SAE/ASEE Joint Propulsion Conference.

[9] Jankovsky, R.S., 1996. HAN-based monopropellant assessment for spacecraft. In: 32nd AIAA/ASME/SAE/ASEE Joint Propulsion Conference and Exhibit.

[10] Freudenmann, D., Ciezki, H.K., 2019. ADN and HAN-based monopropellants — a minireview on compatibility and chemical stability in aqueous media. Propellants, Explos. Pyrotech. 44 (9), 1084—1089.

[11] Chang, Y.P., Reed, B., Josten, K., Kuo, K., 2002. Combustion characteristics of energetic HAN/methanol-based monopropellants. In: 38th AIAA/ASME/SAE/ASEE Joint Propulsion Conference & Exhibit.

[12] Vosen, S.R., 1988. The burning rate of hydroxylammonium nitrate-based liquid propellants. Symp. (Int.) Combust. 22 (1), 1817—1825.

[13] Zhu, D.L., Law, C.K., 1987. Aerothermochemical studies of energetic liquid materials: 1. combustion of HAN-based liquid gun propellants under atmospheric pressure. Combust. Flame 70 (3), 333—342.

[14] Decker, M.M., Freedman, E., Klein, N., Leveritt, C.S., Wojciechowski, J.Q., 1987. HAN-based Liquid Gun Propellants: Physical Properties. US Army Ballistic Research Laboratory, Aberdeen Proving Ground, Maryland.

[15] Knapton, J.D., Morrison, W.F., Klingenberg, G., Wren, G.P. (Eds.), 1998. Liquid Propellant Gun Technology, Progress in Astronautics and Aeronautics. American Institute of Aeronautics and Astronautics, Washington, DC.

[16] Masse, R., Allen, M., Spores, R., Driscoll, E.A., 2016. AF-M315E propulsion system advances and improvements. In: 52nd AIAA/SAE/ASEE Joint Propulsion Conference.

[17] Masse, R., Overly, J., Allen, M., Spores, R., 2012. A new state-of-the-art in AF-M315E thruster technologies. In: 48th AIAA/ASME/SAE/ASEE Joint Propulsion Conference & Exhibit.

[18] Amrousse, R., Katsumi, T., Azuma, N., Hori, K., 2017. Hydroxylammonium nitrate (HAN)-based green propellant as alternative energy resource for potential hydrazine substitution: from lab scale to pilot plant scale-up. Combust. Flame 176, 334—348.

[19] Chai, W.S., Cheah, K.H., Koh, K.S., Chin, J., Chik, T.F.W.K., 2016. Parametric studies of electrolytic decomposition of hydroxylammonium nitrate (HAN) energetic ionic liquid in microreactor using image processing technique. Chem. Eng. J. 296, 19—27.

[20] Chai, W.S., 2017. Characterization & Analysis on Electrolytic Decomposition of Hydroxylammonium Nitrate (HAN) Ternary Mixtures in Microreactors. University of Nottingham.

[21] Kesore, K., Janowski, F., Shaposhnik, V.A., 1997. Highly effective electrodialysis for selective elimination of nitrates from drinking water. J. Membr. Sci. 127 (1), 17—24.

[22] Wheelwright, E.J., 1977. Cation-exchange conversion of hydroxylamine sulfate to hydroxylamine nitrate. Ind. Eng. Chem. Process Des. Dev. 16 (2), 220—222.

[23] Zhao, F., You, K., Li, R., Tan, S., Liu, P., Wu, J., Ai, Q., Luo, H.A., 2015. Novel approach for the preparation of hydroxylammonium nitrate from the acid-catalyzed hydrolysis of cyclohexanone oxime. Ind. Eng. Chem. Res. 54 (3), 819—823.

[24] Koh, K.S., Chin, J., Wahida Ku Chik, T.F., 2013. Role of electrodes in ambient electrolytic decomposition of hydroxylammonium nitrate (HAN) solutions. Propul. Power Res. 2 (3), 194—200.

[25] Vanderhoff, J.A., Bunte, S.W., 1986. Electrical Conductivity Measurements of Hydroxylammonium Nitrate: Design Considerations. US Army Ballistic Research Laboratory, Aberdeen Proving Ground, Maryland.

[26] Schoppelrei, J.W., Brill, T.B., 1997. Spectroscopy of hydrothermal reactions. 7. Kinetics of aqueous [NH3OH]NO3 at 463-523 K and 27.5 MPa by infrared spectroscopy. J. Phys. Chem. 101 (46), 8593—8596.

[27] Kondrikov, B.N., Annikov, V.É., Egorshev, V.Y., De Luca, L.T., 2000. Burning of hydroxylammonium nitrate. Combust. Explos. Shock Waves 36, 135−145.

[28] Van Dijk, C.A., Priest, R.G., 1984. Thermal decomposition of hydroxylammonium nitrate at kilobar pressures. Combust. Flame 57 (1), 15−24.

[29] Farshchi, M., Vaezi, V., Shaw, B.D., 2002. Studies of HAN-based monopropellant droplet combustion. Combust. Sci. Technol. 174 (7), 71−97.

[30] Schoppelrei, J.W., Kieke, M.L., Brill, T.B., 1996. Spectroscopy of hydrothermal reactions. 2. Reactions and kinetic parameters of [NH$_3$OH]NO$_3$ and equilibria of (NH$_4$)$_2$CO$_3$ determined with a flow cell and FT Raman spectroscopy. J. Phys. Chem. 100 (18), 7463−7470.

[31] Klein, N., Wong, K.N., 1987. An Infra-red Investigation of HAN (Hydroxylammonium Nitrate)-Based Liquid Propellants. US Army Ballistic Research Laboratory, Aberdeen Proving Ground, Maryland.

[32] Weller, R.E., Morris, J.E., McClanahan, B.J., Jostes, R.F., Mahlum, D.D., 1989. Evaluation of the Dermal Toxicity of LP1846. Pacific Northwest Laboratory, Richland, Washington, D.C.

[33] Hui, A., Jinyi, L., Lujun, Y., Shengxue, L., Yanhong, Z., Huan, Y., Qingjun, J., Zhihong, C., Jia, C., 2008. Acute and subchronic toxicity of hydroxylammonium nitrate in Wistar rats. J. Med. Coll. PLA 23 (3), 137−147.

[34] Amrousse, R., Kagawa, H., Hatai, K., Ikeda, H., Hori, K., 2015. The effect of iron metal ions and chelating agents of iron on the thermal decomposition of HAN-based liquid monopropellant. In: 51st AIAA/SAE/ASEE Joint Propulsion Conference.

[35] Cronin, J.T., Brill, T.B., 1986. Thermal decomposition of energetic materials. 8. Evidence of an oscillating process during the high-rate thermolysis of hydroxylammonium nitrate, and comments on the interionic interactions. J. Phys. Chem. 90 (1), 178−181.

[36] Klein, N., 1983. Preparation and Characterization of Several Liquid Propellants. US Army Ballistic Research Laboratory, Aberdeen Proving Ground, Maryland.

[37] Vanderhoff, J.A., Bunte, S.W., Donmoyer, P.M., 1987. Laser Raman Studies Related to Liquid Propellants: Structural Characteristics Inferred from the Nitrate Anion Spectra. US Army Ballistic Research Laboratory, Aberdeen Proving Ground, Maryland.

[38] Chai, W.S., Cheah, K.H., Meng, H., Li, G., 2019. Experimental and analytical study on electrolytic decomposition of HAN water solution using graphite electrodes. J. Mol. Liq. 293, 111496.

[39] Harlow, D.G., Felt, R.E., Agnew, S., Barney, G.C., McKibben, J.M., Garber, J., Lewis, M., 1998. Technical Report on Hydroxylamine Nitrate. US Department of Energy.

[40] Wei, C., Rogers, R.J., Mannan, M.S., 2006. Thermal decomposition hazard evaluation of hydroxylamine nitrate. J. Hazard Mater. 130 (1−2), 163−168.

[41] Oxley, J.C., Brower, K.R., 1988. Thermal decomposition of hydroxylamine nitrate. In: 1988 Los Angeles Symposium: O-E/LASE, vol. 88.

[42] Pembridge, J.R., Stedman, G., 1979. Kinetics, mechanism, and stoichiometry of the oxidation of hydroxylamine by nitric acid. J. Chem. Soc., Dalton Trans. 11, 1657−1663.

[43] Rafeev, V.A., Rubtsov, Y.I., 1993. Kinetics and mechanism of thermal decomposition of hydroxylammonium nitrate. Russ. Chem. Bull. 42, 1811−1815.

[44] Zhang, K., Thynell, S.T., 2018. Thermal decomposition mechanism of aqueous hydroxylammonium nitrate (HAN): molecular simulation and kinetic modeling. J. Phys. Chem. 122 (41), 8086−8100.

[45] Liu, L., Papadaki, M., Pontiki, E., Stathi, P., Rogers, W.J., Mannan, M.S., 2009. Isothermal decomposition of hydroxylamine and hydroxylamine nitrate in aqueous solutions in the temperature range 80−160°C. J. Hazard Mater. 165 (1−3), 573−578.

[46] McFarlane, J., Delmau, L.H., DePaoli, D.W., Mattus, C.H., Phelps, C.E., Roach, B.D., 2015. Hydroxylamine Nitrate Decomposition Under Non-radiological Conditions. Oak Ridge National Lab, Oak Ridge, Tennessee.

[47] Liu, J., An, Z., Zhang, Q., Wang, C., 2017. Thermal decomposition of hydroxylamine nitrate studied by differential scanning calorimetry analysis and density functional theory calculations. Prog. React. Kinet. Mech. 42 (4), 334−343.

[48] Nagase, S., Miyazaki, S., Ayabe, M., Kohno, M., 2004. A Preliminary work on HAN/HN-based monopropellant. In: 24th International Symposium on Space Technology and Science.

[49] Azuma, N., Hori, K., Katsumi, T., Amrousse, R., Nagata, T., Hatai, K., Kobayashi, T., Nakayama, Y., Matsumura, T., Fujiwara, S., 2013. Research and development on thrusters with HAN (hydroxyl ammonium nitrate) based monopropellant. In: 5th European Conference for Aerospace Sciences,.

[50] Quach, P., Brand, A., Warmoth, G., 2015. Adiabatic compression testing of AF-M315E. In: 51st AIAA/SAE/ASEE Joint Propulsion Conference.

[51] Meinhardt, D., Christofferson, S., Wucherer, E., Reed, B., 1999. Performance and life testing of small HAN thruster. In: 35th AIAA/SAE/ASEE Joint Propulsion Conference.

[52] Hisatsune, K., Izumi, J., Tsutaya, H., Furukawa, K., 2004. Development of HAN-based liquid properllant thruster. In: 2nd International Conference on Green Propellants for Space Propulsion.

[53] Agnihotri, R., Oommen, C., 2018. Cerium oxide based active catalyst for hydroxylammonium nitrate (HAN) fueled monopropellant thrusters. RSC Adv. 8 (4), 22293−22302.

[54] Lee, H., Litzinger, T.A., 2003. Chemical kinetic study of HAN decomposition. Combust. Flame 135 (1−2), 151−169.

[55] Katsumi, T., Amrousse, R., Niboshi, Y., Hori, K., 2015. A study on the combustion mechanism of hydroxylammonium nitrate. Int. J. Energ. Mater. Chem. Propuls. 14 (4), 307−319.

[56] Esparza, A.A., Ferguson, R.E., Choudhuri, A., Love, N.D., Shafirovich, E., 2018. Thermoanalytical studies on the thermal and catalytic decomposition of aqueous hydroxylammonium nitrate solution. Combust. Flame 193, 417−423.

[57] Amrousse, R., Hori, K., Fetimi, W., Farhat, K., 2012. HAN and ADN as liquid ionic monopropellants: thermal and catalytic decomposition processes. Appl. Catal. B Environ. 127, 121−128.

[58] Shaw, B.D., Williams, F.A., 1992. A model for the deflagration of aqueous solutions of hydroxylammonium nitrate. Symp. (Int.) Combust. 24 (1), 1923−1930.

[59] Chambreau, S.D., Popolan-Vaida, D.M., Vaghjiani, G.L., Leone, S.R., 2017. Catalytic decomposition of hydroxylammonium nitrate ionic liquid: enhancement of NO formation. J. Phys. Chem. Lett. 8 (10), 2126−2130.

[60] Agnihotri, R., Oommen, C., 2021. Kinetics and mechanism of thermal and catalytic decomposition of hydroxylammonium nitrate (HAN) monopropellant, Propellants, Explosives. Pyrotechnics 46 (2), 286−298.

[61] Amrousse, R., Katsumi, T., Bachar, A., Brahmi, B., Bensitel, M., Hori, K., 2014. Chemical engineering study for hydroxylammonium nitrate monopropellant decomposition over monolith and grain metal-based catalysts. React. Kinet. Mech. Catal. 111, 71−88.

[62] Amrousse, R., Katsumi, T., Itouyama, N., Azuma, N., Kagawa, H., Hatai, K., Ikeda, H., Hori, K., 2015. New HAN-based mixtures for reaction control system and low toxic spacecraft propulsion subsystem: thermal decomposition and possible thruster applications. Combust. Flame 162 (6), 2686–2692.

[63] Amrousse, R., Katsumi, T., Niboshi, Y., Azuma, N., Bachar, A., Hori, K., 2013. Performance and deactivation of Ir-based catalyst during hydroxylammonium nitrate catalytic decomposition. Appl. Catal. Gen. 452, 64–68.

[64] Courthéoux, L., Popa, F., Gautron, E., Rossignol, S., Kappenstein, C., 2004. Platinum supported on doped alumina catalysts for propulsion applications. Xerogels versus aerogels. J. Non-Cryst. Solids 350, 113–119.

[65] Ren, X., Li, M., Wang, A., Li, L., Wang, X., Zhang, T., 2007. Catalytic decomposition of hydroxyl ammonium nitrate at room temperature. Chin. J. Catal. 28 (1), 1–2.

[66] Amariei, D., Rossignol, S., Kappenstein, C., Joulin, J.-P., 2006. Shape forming of Pt/Al_2O_3Si sol-gel catalysts for space applications. In: Gaigneaux, E.M., Devillers, M., De Vos, D.E., Hermans, S., Jacobs, P.A., Martens, J.A., Ruiz, P. (Eds.), Studies in Surface Science and Catalysis, Scientific Bases for the Preparation of Heterogeneous Catalysts, vol. 162. Elsevier, pp. 969–976.

[67] Agnihotri, R., Oommen, C., 2020. Evaluation of hydroxylammonium nitrate (HAN) decomposition using bifunctional catalyst for thruster application. Mol. Catal. 486, 110851.

[68] Katsumi, T., Matsuda, R., Inoue, T., Tsuboi, N., Ogawa, H., Sawai, S., Hori, K., 2010. Combustion characteristics of hydroxylammonium nitrate aqueous solutions. Int. J. Energ. Mater. Chem. Propuls. 9 (3), 219–231.

[69] Atamanov, M.K., Amrousse, R., Hori, K., Mansurov, Z., 2019. Experimental investigations of combustion: (95 WT.-%) HAN–water solution with high-SSA activated carbons. Combust. Sci. Technol. 191 (4), 645–658.

[70] Thomas, J.C., Homan-Cruz, G.D., Stahl, J.M., Petersen, E.L., 2019. The effects of SiO_2 and TiO_2 on the two-phase burning behavior of aqueous HAN propellant. Proc. Combust. Inst. 37 (3), 3159–3166.

[71] Amariei, D., Courthéoux, L., Rossignol, S., Kappenstein, C., 2007. Catalytic and thermal decomposition of ionic liquid monopropellants using a dynamic reactor. Comparison of powder and sphere-shaped catalysts. Chem. Eng. Process 46 (2), 165–174.

[72] Fukuchi, B.A., Nagase, S., Maruizumi, H., Ayabe, M., 2010. HAN/HN-based monopropellant thrusters. IHI Eng. Rev. 43 (1), 22–28.

[73] Eloirdi, R., Rossignol, S., Kappenstein, C., Duprez, D., Pillet, N., 2003. Design and use of a batch reactor for catalytic decomposition of propellants. J. Propul. Power 19, 213–219.

Index

Printed in the United States
by Baker & Taylor Publisher Services